PHILIP K

PETER GU

INTRODUCTION

TO

QUATERNIONS,

WITH NUMEROUS EXAMPLES

Elibron Classics

www.elibron.com

Elibron Classics series.

© 2005 Adamant Media Corporation.

ISBN 1-4021-0069-8 (paperback)
ISBN 1-4021-5410-0 (hardcover)

This Elibron Classics Replica Edition is an unabridged facsimile
of the edition published in 1882 by Macmillan and Co.,
London.

INTRODUCTION

TO

QUATERNIONS,

WITH NUMEROUS EXAMPLES.

INTRODUCTION

TO

QUATERNIONS,

WITH NUMEROUS EXAMPLES.

BY

P. KELLAND, M.A., F.R.S.,

FORMERLY FELLOW OF QUEENS' COLLEGE, CAMBRIDGE;

AND

P. G. TAIT, M.A., Sec. R.S.E.,

FORMERLY FELLOW OF ST PETER'S COLLEGE, CAMBRIDGE;

PROFESSORS IN THE DEPARTMENT OF MATHEMATICS IN THE UNIVERSITY OF EDINBURGH.

SECOND EDITION.

London:

MACMILLAN AND CO.

1882

𝕮𝖆𝖒𝖇𝖗𝖎𝖉𝖌𝖊 :

PRINTED BY C. J. CLAY, M.A.
AT THE UNIVERSITY PRESS.

PREFACE.

In preparing this second edition for press I have altered as slightly as possible those portions of the work which were written entirely by Prof. Kelland. The mode of presentation which he employed must always be of great interest, if only from the fact that he was an exceptionally able teacher; but the success of the work, as an introduction to a method which is now rapidly advancing in general estimation, would of itself have been a sufficient motive for my refraining from any serious alteration.

A third reason, had such been necessary, would have presented itself in the fact that I have never considered with the necessary care those metaphysical questions connected with the growth and development of mathematical ideas, to which my late venerated teacher paid such particular attention.

My own part of the book (including mainly Chap. X. and worked out Examples 10—24 in Chap. IX.) was written hurriedly, and while I was deeply engaged with work of a very different kind; so that I had no hesitation in determining to re-cast it where I fancied I could improve it.

P. G. TAIT.

University of Edinburgh,
November, 1881.

PREFACE TO THE FIRST EDITION.

THE present Treatise is, as the title-page indicates, the joint production of Prof. Tait and myself. The preface I write in the first person, as this enables me to offer some personal explanations.

For many years past I have been accustomed, no doubt very imperfectly, to introduce to my class the subject of Quaternions as part of elementary Algebra, more with the view of establishing principles than of applying processes. Experience has taught me that to induce a student to think for himself there is nothing so effectual as to lay before him the different stages of the development of a science in something like the historical order. And justice alike to the student and the subject forbade that I should stop short at that point where, more simply and more effectually than at any other, the intimate connexion between principles and processes is made manifest. Moreover, in lecturing on the groundwork on which the mathematical sciences are based, I could not but bring before my class the names of great men who spoke in other tongues and belonged to other nationalities than their own—Diophantus, Des Cartes, Lagrange, for instance—and it was not just to omit the name of one as

great as any of them, Sir William Rowan Hamilton, who
spoke their own tongue and claimed their own nationality.
It is true the name of Hamilton has not had the impress
of time to stamp it with the seal of immortality. And it
must be admitted that a cautious policy which forbids to
wander from the beaten paths, and encourages converse
with the past rather than interference with the present, is
the true policy of a teacher. But in the case before us,
quite irrespective of the nationality of the inventor, there
is ample ground for introducing this subject of Quaternions
into an elementary course of mathematics. It belongs to
first principles and is their crowning and completion. It
brings those principles face to face with operations, and thus
not only satisfies the student of the mutual dependence of
the two, but tends to carry him back to a clear apprehension
of what he had probably failed to appreciate in the sub-
ordinate sciences.

Besides, there is no branch of mathematics in which
results of such wide variety are deduced by one uniform
process; there is no territory like this to be attacked
and subjugated by a single weapon. And what is of the
utmost importance in an educational point of view, the
reader of this subject does not require to encumber his
memory with a host of conclusions already arrived at in
order to advance. Every problem is more or less self-
contained. This is my apology for the present treatise.

The work is, as I have said, the joint production
of Prof. Tait and myself. The preface I have written
without consulting my colleague, as I am thus enabled

to say what could not otherwise have been said, that mathematicians owe a lasting debt of gratitude to Prof. Tait for the singleness of purpose and the self-denying zeal with which he has worked out the designs of his friend Sir Wm. Hamilton, preferring always the claims of the science and of its founder to the assertion of his own power and originality in its development. For my own part I must confess that my knowledge of Quaternions is due exclusively to him. The first work of Sir Wm. Hamilton, *Lectures on Quaternions*, was very dimly and imperfectly understood by me and I dare say by others, until Prof. Tait published his papers on the subject in the *Messenger of Mathematics*. Then, and not till then, did the science in all its simplicity develope itself to me. Subsequently Prof. Tait has published a work of great value and originality, *An Elementary Treatise on Quaternions*.

The literature of the subject is completed in all but what relates to its physical applications, when I mention in addition Hamilton's second great work, *Elements of Quaternions*, a posthumous work so far as publication is concerned, but one of which the sheets had been corrected by the author, and which bears all the impress of his genius. But it is far from elementary, whatever its title may seem to imply; nor is the work of Prof. Tait altogether free from difficulties. Hamilton and Tait write for mathematicians, and they do well, but the time has come when it behoves some one to write for those who desire to become mathematicians. Friends and pupils have urged me to undertake this duty, and after consultation with Prof. Tait, who from being my pupil in youth is my teacher in riper years, I have, in conjunction with him, and drawing unreservedly from his writings, endeavoured in the first nine chapters of this treatise to illustrate and enforce the principles of this beautiful science. The last chapter, which may be regarded as an introduction to the application of Quaternions to the region beyond that of pure geometry, is due to Prof. Tait alone. Sir W. Hamilton, on nearly the last completed page of his last work, indicated Prof. Tait as eminently fitted to carry on happily and usefully the applications, mathematical and physical, of Quaternions, and as likely to become in the science one of the chief successors of its inventor. With how great justice, the reader of this chapter and of Prof. Tait's other writings on the subject will judge.

PHILIP KELLAND.

University of Edinburgh,
   October, 1873.

CONTENTS.

INTRODUCTION TO QUATERNIONS.

CHAPTER I.

THE science named Quaternions by its illustrious founder, Sir William Rowan Hamilton, is the last and the most beautiful example of extension by the removal of limitations.

The Algebraic sciences are based on ordinary arithmetic, starting at first with all its restrictions, but gradually freeing themselves from one and another, until the parent science scarce recognises itself in its offspring. A student will best get an idea of the thing by considering one case of extension within the science of Arithmetic itself. There are two distinct bases of operation in that science—addition and multiplication. In the infancy of the science the latter was a mere repetition of the former. Multiplication was, in fact, an abbreviated form of equal additions. It is in this form that it occurs in the earliest writer on arithmetic whose works have come down to us—Euclid. Within the limits to which his principles extended, the reasonings and conclusions of Euclid in his seventh and following Books are absolutely perfect. The demonstration of the rule for finding the greatest common measure of two numbers in Prop. 2, Book VII. is identically the same as that which is given in all modern treatises. But Euclid dares not venture on fractions. Their properties were probably all but unknown to him. Accordingly we look in vain for any *demonstration* of the properties of fractions in the writings of the Greek arithmeticians. For that we must come lower down. On the revival of science in the West, we are presented with categorical treatises on arithmetic. The first printed treatise is that of Lucas de Burgo in 1494. The author considers a fraction to be a quotient, and thus, as he expressly states, the order of operations becomes the reverse of that for whole numbers—multiplication precedes addition, etc. In our own country we have a tolerably early writer on arithmetic, Robert Record, who dedicated his work to King Edward the Sixth. The ingenious author exhibits his treatise in the form of a dialogue between master and scholar. The scholar battles long with this difficulty—that multiplying a thing should make it less. At first, the master attempts to explain the anomaly by reference to proportion, thus : that the product by a fraction bears the same proportion to the thing multiplied that the multiplying fraction does to unity. The scholar is not satisfied ; and accordingly the master goes on to say : "If I multiply by more than one, the thing is increased ; if I take it but once, it is not changed; and if I take it less than once, it cannot be so much as it was before. Then, seeing that a fraction is less than one, if I multiply by a fraction, it follows that I do take it less than once", etc. The scholar thereupon replies, " Sir, I do thank you much for this reason ; and I trust that I do perceive the thing".

Need we add that the same difficulty which the scholar in the time of King Edward experienced, is experienced by every thinking boy of our own times; and the explanation afforded him is precisely the same admixture of multiplication, proportion, and division which suggested itself to old Robert Record. Every schoolboy feels that to multiply by a fraction is not to multiply at all in the sense in which multiplication was originally presented to him, viz. as an abbreviation of equal additions, or of repetitions of the thing multiplied. A totally new view of the process of multiplication has insensibly crept in by the advance from whole numbers to fractions. So new, so different is it, that we are satisfied Euclid in his logical and unbending march could never have attained to it. It is only by standing loose for a time to logical accuracy that extensions in the abstract sciences—extensions at any rate which stretch from one science to another—are effected. Thus Diophantus in his

Treatise on Arithmetic (i. e. Arithmetic extended to Algebra) boldly lays it down as a definition or first principle of his science that 'minus into minus makes plus'. The science he is founding is subject to this condition, and the results must be interpreted consistently with it. So far as this condition does not belong to ordinary arithmetic, so far the science extends beyond ordinary arithmetic: and this is the distance to which it extends—It makes subtraction to stand by itself, apart from addition; or, at any rate, not dependent on it.

We trust, then, it begins to be seen that sciences are extended by the removal of barriers, of limitations, of conditions, on which sometimes their very existence appears to depend. Fractional arithmetic was an impossibility so long as multiplication was regarded as abbreviated addition; the moment an extended idea was entertained, ever so illogically, that moment fractional arithmetic started into existence. Algebra, except as mere symbolized arithmetic, was an impossibility so long as the thought of subtraction was chained to the requirement of something adequate to subtract from. The moment Diophantus gave it a separate existence—boldly and logically as it happened—by exhibiting the law of *minus* in the forefront as the primary definition of his science, that moment algebra in its highest form became a possibility; and indeed the foundation-stone was no sooner laid than a goodly building arose on it.

The examples we have given, perhaps from their very simplicity, escape notice, but they are not less really examples of extension from science to science by the removal of a restriction. We have selected them in preference to the more familiar one of the extension of the meaning of an index, whereby it becomes a logarithm, because they prepare the way for a further extension in the same direction to which we are presently to advance. Observe, then, that in fractions and in the rule of signs, addition (or subtraction) is very slenderly connected with multiplication (or division). Arithmetic as Euclid left it stands on one support, addition only, inasmuch as with him multiplication is but abbreviated addition. Arithmetic in its extended form rests on two supports, addition and multiplication, the one different from the other. This is the first idea we want our reader to get a firm hold of; that multiplication is not necessarily addition, but an operation self-contained, self-interpretable—springing originally out of addition; but, when full-grown, existing apart from its parent.

The second idea we want our reader to fix his mind on is this, that when a science has been extended into a new form, certain limitations, which appeared to be of the nature of essential truths in the old science, are found to be utterly untenable; that it is, in fact, by throwing these limitations aside that room is made for the growth of the new science. We have instanced Algebra as a growth out of Arithmetic by the removal of the restriction that subtraction shall require something to subtract from. The word 'subtraction' may indeed be inappropriate, as the word multiplication appeared to be to Record's scholar, who failed to see how the multiplication of a thing could make it less. In the advance of the sciences the old terminology often becomes inappropriate; but if the mind can extract the right idea from the sound or sight of a word, it is the part of wisdom to retain it. And so all the old words have been retained in the science of Quaternions to which we are now to advance.

The fundamental idea on which the science is based is that of motion—of transference. Real motion is indeed not needed, any more than real superposition is needed in Euclid's Geometry. An appeal is made to mental transference in the one science, to mental superposition in the other.

We are then to consider how it is possible to frame a new science which shall spring out of Arithmetic, Algebra, and Geometry, and shall add to them the idea of motion—of transference. It must be confessed the project we entertain is not a project due to the nineteenth century. The Geometry of Des Cartes was based on something very much resembling the idea of motion, and so far the mere introduction of the idea of transference was not of much value. The real advance was due to the thought of severing multiplication from addition, so that the one might be the representative of a kind of motion absolutely different from that which was represented by the other, yet capable of being combined with it. What the nine-teenth century has done, then, is to divorce addition from multipli-cation in the new form in which the two are presented, and to cause the one, in this new character, to signify motion forwards and backwards, the other motion round and round.

We do not purpose to give a history of the science, and shall accordingly content ourselves with saying, that the notion of sepa-rating addition from multiplication—attributing to the one, motion from a point, to the other motion about a point—had been floating in the minds of mathematicians for half a century, without producing many results worth recording, when the subject fell into the hands of a giant, Sir William Rowan Hamilton, who early found that his road was obstructed—he knew not by what obstacle—so that many points which seemed within his reach were really inaccessible. He had done a considerable amount of good work, obstructed as he was, when, about the year 1843, he perceived clearly the obstruction to his progress in the shape of an old law which, prior to that time, had appeared like a law of common sense. The law in question is known as the *commutative* law of multiplication. Presented in its simplest form it is nothing more than this, 'five times three is the same as three times five'; more generally, it appears under the form of '$ab = ba$ whatever a and b may represent'. When it came distinctly into the mind of Hamilton that this law is not a necessity, with the extended signification of multiplication, he saw his way clear, and gave up the law. The barrier being removed, he entered on the new science as a warrior enters a besieged city through a practicable breach. The reader will find it easy to enter after him.

CHAPTER II.

VECTOR ADDITION AND SUBTRACTION.

1. *Definition of a Vector.* A vector is the representative of transference through a given distance, in a given direction. Thus if AB be a straight line, the idea to be attached to 'vector AB' is that of transference from A to B.

For the sake of definiteness we shall frequently abbreviate the phrase 'vector AB' by a Greek letter, retaining in the meantime (with one exception to be noted in the next chapter) the English letters to denote ordinary numerical quantities.

If we now start from B and advance to C in the same direction, BC being equal to AB, we may, as in ordinary geometry, designate 'vector BC' by the same symbol, which we adopted to designate 'vector AB.'

Further, if we start from any other point O in space, and advance from that point by the distance OX equal to and in the same direction as AB, we are at liberty to designate 'vector OX' by the same symbol as that which represents AB.

Other circumstances will determine the starting point, and individualize the line to which a specific vector corresponds. Our definition is therefore subject to the following condition :—*All lines which are equal and drawn in the same direction are represented by the same vector symbol.*

We have purposely employed the phrase 'drawn in the same direction' instead of 'parallel,' because we wish to guard the student against confounding 'vector AB' with 'vector BA.'

2. In order to apply algebra to geometry, it is necessary to impose on geometry the condition that when a line measured in one direction is represented by a *positive* symbol, the same line measured in the opposite direction must be represented by the corresponding *negative* symbol.

In the science before us the same condition is equally requisite, and indeed the reason for it is even more manifest. For if a transference from A to B be represented by $+a$, the transference which neutralizes this, and brings us back again to A, cannot be conceived to be represented by anything but $-a$, provided the symbols $+$ and $-$ are to retain any of their old algebraic meaning. The vector AB, then, being represented by $+a$, the vector BA will be represented by $-a$.

3. Further it is abundantly evident that so far as addition and subtraction of parallel vectors are concerned, *all* the laws of Algebra must be applicable. Thus (in Art. 1) $AB + BC$ or $a + a$ produces the same result as AC which is twice as great as AB, and is therefore properly represented by $2a$; and so on for all the rest. The *distributive* law of addition may then be assumed to hold in all its integrity so long at least as we deal with vectors which are parallel to one another. In fact there is no reason whatever, so far, why a should not be treated in every respect as if it were an ordinary algebraic quantity. It need scarcely be added that vectors in the same direction have the same proportion as the *lines* which correspond to them.

We have then advanced to the following—

LEMMA. *All lines drawn in the same direction are, as vectors, to be represented by numerical multiples of one and the same symbol, to which the ordinary laws of Algebra, so far as their addition, subtraction, and numerical multiplication are concerned, may be unreservedly applied.*

4. The converse is of course true, that if lines as vectors are represented by multiples of the same vector symbol, they are parallel.

It is only necessary to add to what has preceded, that if BC be a line *not* in the same direction with AB, then the vector BC cannot be represented by α or by any (arithmetical) multiple of α. The vector symbol α must be limited to express transference in a certain direction, and cannot, at the same time, express transference in any other direction. To express 'vector BC' then, another and quite independent symbol β must be introduced. This symbol, being united to α by the signs + and −, the laws of algebra will, of course, apply to the combination.

5. If we now join AC, and thus form a triangle ABC, and if we denote vector AB by α, BC by β, AC by γ, it is clear that we shall be presented with the equation $\alpha + \beta = \gamma$.

This equation appears at first sight to be a violation of Euclid I. 20 : "Any two sides of a triangle are together greater than the third side". But it is not really so. The anomalous appearance arises from the fact that whilst we have extended the meaning of the symbol + beyond its arithmetical signification, we have said nothing about that of a symbol =. It is clearly necessary that the signification of this symbol shall be extended along with that of the other. It must now be held to designate, as it does perpetually in algebra, 'equivalent to.' This being premised, the equation above is freed from its anomalous appearance, and is perfectly consistent with everything in ordinary geometry. Expressed in words it reads thus : 'A transference from A to B followed by a transference from B to C is equivalent to a transference from A to C.'

6. AXIOM. *If two vectors have not the same direction, it is impossible that the one can neutralize the other.*

This is quite obvious, for when a transference has been effected from A to B, it is impossible to conceive that any amount of transference whatever along BC can bring the moving point back to A.

It follows as a consequence of this axiom, that if α, β be *different* actual vectors, i.e. finite vectors not in the same direction, and if

$ma + n\beta = 0$, where m and n are numerical quantities; then must $m = 0$ and $n = 0$.

Another form of this consequence may be thus stated. If [still with the above assumption as to α and β] $ma + n\beta = pa + q\beta$, then must $m = p$, and $n = q$.

7. We now proceed to exemplify the principles so far as they have hitherto been laid down. It is scarcely necessary to remind the reader that we are assuming the applicability of all the rules of algebra and arithmetic, so far as we are yet in a position to draw on them; and consequently that our demonstrations of certain of Euclid's elementary propositions must be accepted subject to this assumption.

To avoid prolixity, we shall very frequently drop the word *vector*, at least in cases where, either from the introduction of a Greek letter as its representative, or from obvious considerations, it must be clear that the mere line is not meant. The reader will not fail to notice that the method of demonstration consists mainly in reaching the same point by two different routes. (See remark on Ex. 9.)

EXAMPLES.

Ex. 1. *The straight lines which join the extremities of equal and parallel straight lines towards the same parts are themselves equal and parallel.*

Let AB be equal and parallel to CD; to prove that AC is equal and parallel to BD.

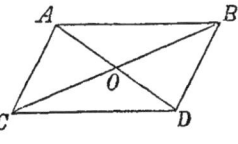

Let vector AB be represented by α, then (Art. 1) vector CD is also represented by α.

If now vector CA be represented by β, vector DB by γ, we shall have (Art. 5)   vector $CB = CA + AB = \beta + \alpha$,

$$\text{and vector } CB = CD + DB = \alpha + \gamma;$$
$$\therefore \quad \beta + \alpha = \alpha + \gamma,$$
$$\text{and } \beta = \gamma;$$

so that β and γ are the same vector symbol; consequently (Art. 1) the *lines* which they represent are equal and parallel; i.e. CA is equal and parallel to BD.

Ex. 2.   *The opposite sides of a parallelogram are equal; and the diagonals bisect each other.*

Since AB is parallel to CD, if vector AB be represented by a, vector CD will be represented by some numerical multiple of a (Art. 3), call it ma.

And since CA is parallel to DB; if vector CA be β, then vector DB is $n\beta$; hence

$$\text{vector } CB = CA + AB = \beta + a,$$
$$\text{and } = CD + DB = ma + n\beta;$$
$$\therefore \ a + \beta = ma + n\beta.$$

Hence (Art. 6) $m = 1$, $n = 1$, i.e. the opposite sides of the parallelogram are equal.

Again, as vectors, $AO + OB = AB$
$$= CD$$
$$= CO + OD;$$

And as AO is a vector *along* OD, and CO a vector *along* OB; it follows (Art. 6) that vector AO *is* vector OD, and vector CO is OB;

$$\therefore \ \text{line } AO = OD, \quad CO = OB.$$

Ex. 3.   *The sides about the equal angles of equiangular triangles are proportionals.*

Let the triangles ABC, ADE have a common angle A, then, because the angles D and B are equal, DE is parallel to BC.

Let vector AD be represented by a, DE by β, then (Art. 3) AB is ma, BC $n\beta$.

$$\therefore \ \text{as vectors, } AE = AD + DE = a + \beta,$$
$$AC = AB + BC = ma + n\beta.$$

Now AC is a multiple of AE, call it $p(a+\beta)$.

$$\therefore \ ma + n\beta = p(a + \beta),$$
$$\text{and } m = p = n \ (\text{Art. 6}).$$

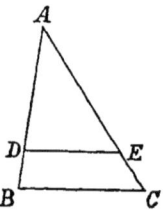

But line $AB : AD = m$,

   line $BC : DE = n$,

   $\therefore AB : AD :: BC : DE$.

Ex. 4. *The bisectors of the sides of a triangle meet in a point which trisects each of them.*

Let the sides of the triangle ABC be bisected in D, E, F; and let AD, BE meet in G.

Let vector BD or DC be α, CE or EA β, then, as vectors,

$BA = BC + CA = 2\alpha + 2\beta = 2(\alpha + \beta)$,

$DE = DC + CE = \alpha + \beta$,

hence (Art. 4) BA is parallel to DE, and equal to $2DE$.

Again, $BG + GA = BA$

$= 2DE$

$= 2(DG + GE)$.

Now vector BG is along GE, and vector GA along DG.

$\therefore$ (Art. 6) $BG = 2GE$,

$GA = 2DG$,

whence the same is true of the lines.

Lastly, $BG = \dfrac{2}{3}BE$

$= \dfrac{2}{3}(BC + CE)$

$= \dfrac{2}{3}(2\alpha + \beta)$;

$\therefore CG = BG - BC$

$= \dfrac{2}{3}(2\alpha + \beta) - 2\alpha$

$= \dfrac{2}{3}(\beta - \alpha)$,

$$GF = BF - BG$$

$$= \frac{1}{2} BA - BG,$$

$$= a + \beta - \frac{2}{3} (2a + \beta)$$

$$= \frac{1}{3} (\beta - a);$$

hence CG is in the same straight line with GF, and equal to $2GF$.

Ex. 5. *When, instead of D and E being the middle points of the sides, they are any points whatever in those sides, it is required to find G and the point in which CG produced meets AB.*

Let $\dfrac{BC}{DC} = m$, $\dfrac{CA}{CE} = n$; also let vector $DC = a$, vector $CE = \beta$;

$$\therefore BC = ma, \quad CA = n\beta.$$

Hence $BE = BC + CE = ma + \beta,$

$$DA = a + n\beta.$$

Let $BG = xBE, \quad GA = yDA,$

then $BA = BG + GA = x (ma + \beta) + y (a + n\beta).$

But $BA = ma + n\beta,$

$\therefore$ (Art. 6) $xm + y = m, \quad x + yn = n,$

and x, i.e. $\dfrac{BG}{BE} = \dfrac{(m-1)\,n}{mn-1}$, y or $\dfrac{AG}{AD} = \dfrac{(n-1)\,m}{mn-1}$.

Again, let $BF = pBA = p (ma + n\beta).$

But $BF = BC + CF$

$$= ma + \text{a multiple of } CG$$

$$= ma + zCG \text{ suppose}$$

$$= ma + z \{BG - BC\}$$

$$= ma + z \left\{ \frac{(m-1)\,n}{mn-1} (ma + \beta) - ma \right\}.$$

The two values of BF being equated, and Art. 6 applied, there results

$$p = 1 - z \frac{n-1}{mn-1}, \quad p = z \frac{m-1}{mn-1},$$

whence
$$\frac{1-p}{p} = \frac{n-1}{m-1};$$

$$\text{i. e. } \frac{AF}{BF} = \frac{AE}{CE} \div \frac{BD}{CD},$$

$$\text{or } AF . BD . CE = AE . CD . BF.$$

Ex. 6.   *When, instead of as in Ex. 4, where D, E, F are points taken within BC, CA, AB at distances equal to half those lines respectively, they are points taken in BC, CA, AB produced, at the same distances respectively from C, A, and B; to find the intersections.*

Let the points of intersection be respectively G_1, G_2, G_3.

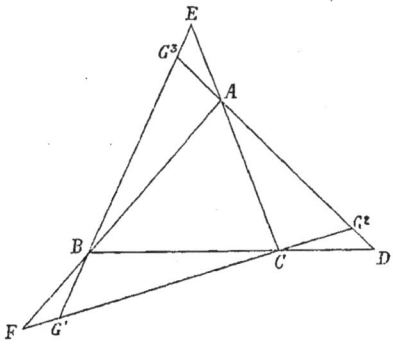

Retaining the notation of Ex. 4, we have

$$BD = 3a, \ CE = 3\beta;$$

$$\text{and } \therefore BG_3 = xBE$$

$$= x(2a + 3\beta) \quad \dots\dots\dots\dots\dots\dots\dots(1),$$

and
$$BG_3 = BD + DG_3$$

$$= 3a + yDA$$

$$= 3a + y(CA - CD)$$

$$= 3a + y(2\beta - a);$$

$$\therefore 2x = 3 - y, \ 3x = 2y, \text{ and } x = \frac{6}{7};$$

$$\therefore \text{ line } EG_3 = \frac{1}{7}EB.$$

Similarly line $FG_1 = \dfrac{1}{7} FC$,

line $DG_2 = \dfrac{1}{7} DA$,

and from equation (1) $BG_3 = \dfrac{6}{7}(2\alpha + 3\beta)$.

But $\qquad BG_3 = BA + AG_3 = 2\alpha + 2\beta + AG_3$;

$$\therefore AG_3 = \dfrac{2}{7}(2\beta - \alpha) ;$$

hence $\qquad$ line $AG_3 = \dfrac{2}{7}$ line DA

$$= 2DG_2,$$

and similarly of the others.

Ex. 7. *The middle points of the lines which join the points of bisection of the opposite sides of a quadrilateral coincide, whether the four sides of the quadrilateral be in the same plane or not.*

Let $ABCD$ be a quadrilateral; E, H, G, F the middle points of AB, BC, CD, DA; X the middle point of EG.

Let vector $AB = \alpha$, $AC = \beta$, $AD = \gamma$,

then $AE + EG = AD + DG$ gives

$$\dfrac{1}{2}\alpha + EG = \gamma + \dfrac{1}{2}(\beta - \gamma),$$

and $AX = AE + \dfrac{1}{2} EG$

$$= \dfrac{1}{4}(\alpha + \beta + \gamma),$$

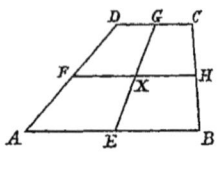

which being symmetrical is α, β, γ in the same as the vector to the middle point of HF.

X is called (Art. 14) the mean point of $ABCD$.

Ex. 8. *The point of bisection of the line which joins the middle points of the diagonals of a quadrilateral (plane or not) is the mean point.*

Let P, Q be the middle points of AC, BD, R that of PQ.

Retaining the notation of the last example we have

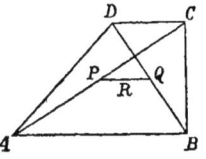

$$AP = \frac{1}{2}\beta,$$

$$AQ = AB + PQ = a + \frac{1}{2}(\gamma - a) = \frac{1}{2}(a + \gamma),$$

$$\text{i.e. } AQ = \frac{1}{2}(AB + AD).$$

Similarly

$$AR = \frac{1}{2}(AP + AQ)$$

$$= \frac{1}{4}(a + \beta + \gamma),$$

i.e. R is the same point as X in the last example; and is therefore the mean point of $ABCD$.

Ex. 9. *AD is drawn bisecting BC in D and is produced to any point E; AB, CE produced meet in P; AC, BE in Q; PQ is parallel to BC.*

Let $AB = a$, $AC = \beta$,

$AP = xa$, $AQ = y\beta$,

$\therefore BC = \beta - a$, $AD = AB + \frac{1}{2}BC$,

$$= \frac{1}{2}(a + \beta)$$

and AE is a multiple of $AD = z(a + \beta)$ say.

Then $CP = pCE$ gives $xa - \beta = p\{z(a + \beta) - \beta\}$,

$$\therefore \text{(Art. 6) } x = pz,\ -1 = pz - p;$$

$$\therefore p = x + 1.$$

Similarly $BQ = qBE$ gives $y\beta - a = q\{z(a + \beta) - a\}$,

$$y = qz,\ -1 = qz - q,$$

$$\therefore q = y + 1,$$

and since $z = \dfrac{x}{p} = \dfrac{y}{q}$ we have

$$x = y, \quad p = q;$$

$$\therefore \ PQ = y\beta - x\alpha = x(\beta - \alpha) = xBC,$$

hence the line PQ is parallel to BC.

The method pursued in this example leads to the solution of all similar problems. It consists, as we have already stated, in reaching the points P and Q respectively by two different routes,—viz. through C and through E for P; through B and through E for Q —and comparing the results.

Cor. 1. $PE : EC :: p-1 : 1 :: x : 1 :: AP : AB.$

Cor. 2. $AE : AD :: 2z : 1 :: 2x : x+1$

$$:: 2(p-1) : p$$

$$:: 2PE : PC,$$

$$\therefore \ AD : DE :: PE + EC : PE - EC.$$

Ex. 10.   *If DEF be drawn cutting the sides of a triangle ; then will* $AD . BF . CE = AE . CF . BD.$

Let $BD = \alpha,\ DA = p\alpha,\ AE = \beta,\ EC = q\beta,$
then $BC = BA + AC = (1+p)\,\alpha + (1+q)\,\beta,$
and CF is a multiple of BC.

Let $CF = xBC$

$$= x\{(1+p)\,\alpha + (1+q)\,\beta\}.$$

But $CF = CE + EF$

$$= -EC + EF$$

$$= -q\beta + y(p\alpha + \beta);$$

$\therefore$ equating, we have $x(1+p) = yp,\ x(1+q) = -q+y,$

whence $x = (1+x)\,pq,$

i. e. $\dfrac{CF}{BC} = \dfrac{BF}{BC} \cdot \dfrac{AD}{BD} \cdot \dfrac{CE}{AE};$

$$\therefore \ AD . BF . CE = AE . CF . BD.$$

Ex. 11. *If from any point within a parallelogram, parallels be drawn to the sides, the corresponding diagonals of the two*

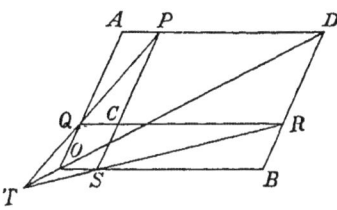

parallelograms thus formed, and of the original parallelogram shall meet in the same point.

Let PQ, RS meet in T;

join TO, OD.

Let $\quad OA = a$, $\ OB = \beta$, $\ OQ = m a$, $\ OS = n \beta$,

then $QP = QC + CP = n\beta + (1 - m)\, a$, $SR = SC + CR = ma + (1-n)\,\beta$,

and $\qquad TO = TQ - OQ = x\left\{n\beta + (1-m)\, a\right\} - ma$,

also $\qquad TO = TS - OS = y\left\{ma + (1-n)\,\beta\right\} - n\beta$:

equating, there results

$$xn = y\,(1-n) - n\,;\ \ x\,(1-m) - m = ym\,;$$

$$\therefore\ x = \frac{m}{1-m-n},$$

and $\qquad TO = \dfrac{mn}{1-m-n}\,(a + \beta) = \dfrac{mn}{1-m-n}\,OD\,;$

hence (Art. 4) TO, OD are in the same straight line.

Cor. $TO : TD :: mn : (1-m)(1-n) :: OSCQ : CRDP.$

Ex. 12. *The points of bisection of the three diagonals of a complete quadrilateral are in a straight line.*

T. Q.   2.

P, Q, R, the middle points of the diagonals of the complete quadrilateral $ABCD$, are in a straight line.

Let $\quad AB = a, \; AD = \beta,$

$\qquad AE = ma, \; AF = n\beta \;$;

$\therefore \; BF = n\beta - a$ and $BC = x\,(n\beta - a)$,

$\quad ED = \beta - ma$ and $CD = y\,(\beta - ma)$.

Now $BC + CD = BD = AD - AB$

gives $\qquad x\,(n\beta - a) + y\,(\beta - ma) = \beta - a$,

whence $\qquad xn + y = 1, \; x + my = 1,$

$$\therefore \; x = \frac{m-1}{mn-1},$$

and
$$AP = \frac{1}{2} AC = \frac{1}{2} \left\{ a + \frac{m-1}{mn-1}\,(n\beta - a) \right\}$$

$$= \frac{1}{2} \frac{m\,(n-1)\,a + n\,(m-1)\,\beta}{mn-1},$$

$$AQ = \frac{1}{2}\,(a + \beta),$$

$$AR = \frac{1}{2}\,(ma + n\beta),$$

$$\therefore \; AQ - AP = \frac{1}{2\,(mn-1)} \{(m-1)\,a + (n-1)\,\beta\},$$

$$AR - AP = \frac{mn}{2\,(mn-1)} \{(m-1)\,a + (n-1)\,\beta\},$$

or vector PR is a multiple of vector PQ, and therefore they are in the same straight line.

Cor.   Line $PQ : PR :: 1 : mn$

$\qquad\qquad :: AB \cdot AD : AE \cdot AF$

$\qquad\qquad ::$ triangle ABD : triangle AEF.

We shall presently exemplify a very elegant method due to Sir W. Hamilton of proving three points to be in the same straight line.

8. It is often convenient to take a vector of the length of the unit, and to express the vector under consideration as a numerical multiple of this unit. Of course it is not necessary that the unit should have any specified value; all that is required is that when once assumed for any given problem, it must remain unchanged throughout the discussion of that problem.

If the line AB be supposed to be a units in length, and the *unit vector* along AB be designated by a, then will *vector AB* be aa (Art. 3).

Sir William Hamilton has termed the length of the line in such cases, the TENSOR of the vector; so that the vector AB is the product of the tensor AB and the unit vector along AB. Thus if, as in the examples worked under the last article, we designate the vector AB by a, we may write $a = TaUa$, where Ta is an abbreviation for 'Tensor of the vector a'; Ua for 'unit vector along a'.

EXAMPLES.

Ex. 1. *If the vertical angle of a triangle be bisected by a straight line which also cuts the base, the segments at the base shall have the same ratio that the other sides of the triangle have to one another.*

Take unit vectors along AB, AC, which call a, β respectively: construct a rhombus $APQR$ on them and draw its diagonal AR. Then since the diagonals of a rhombus bisect its angles, it is clear that the vector 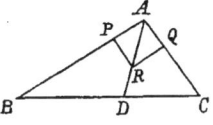 AD which bisects the angle A is a multiple of AR the diagonal vector of the rhombus.

$$\text{Now} \qquad AR = AP + PR = AP + AQ = a + \beta,$$
$$\therefore\ AD = x\,(a + \beta).$$

Now vector $AB = ca$, $AC = b\beta$; using c, b as in ordinary geometry for the lengths of AB, AC.

Hence $\qquad BD = AD - AB = x\,(a + \beta) - ca,$

and $\qquad\qquad BD = yBC = y\,(AC - AB)$
$$= y\,(b\beta - ca).$$

2—2

Equating, $x - c = -yc, \ x = yb$;

$$\therefore \ y = \frac{c}{b+c},$$

and $BD : DC :: y : 1 - y$

$$:: \ c \ : \ b$$

$$:: \ BA \ : \ AC.$$

COR. If α, β are unit vectors from A, and if δ be another vector from A such that $\delta = x(\alpha + \beta)$; then δ bisects the angle between α and β.

EX. 2. *The three bisectors of the angles of a triangle meet in a point.*

Let AD, BE bisect A, B and meet in G; CG bisects C.

Let units along AB, AC, BC be α, β, γ, then as in the last example,

$$AG = x(\alpha + \beta), \ BG = y(-\alpha + \gamma).$$

But $\alpha\gamma = b\beta - c\alpha$,

$$\therefore \ BG = y\left(-\alpha + \frac{b\beta - c\alpha}{a}\right),$$

and $CG = AG - AC$

$$= x(\alpha + \beta) - b\beta,$$

also $CG = BG - BC$

$$= y\left(-\alpha + \frac{b\beta - c\alpha}{a}\right) - b\beta + c\alpha;$$

$$\therefore \ x = -y - \frac{c}{a}y + c,$$

$$x - b = \frac{yb}{a} - b,$$

whence $x = \dfrac{bc}{a+b+c}$,

and $CG = \dfrac{b}{a+b+c}\{c\alpha - (a+b)\beta\}$

$$= \frac{b}{a+b+c}(-\alpha\gamma - a\beta)$$

$$= p(\gamma + \beta),$$

hence CG bisects the angle C (Cor. Ex. 1).

9. If a, β, γ are non-parallel vectors in the same plane, it is always possible to find numerical values of a, b, c so that $aa + b\beta + c\gamma$ shall $= 0$.

For a triangle can be constructed whose sides shall be parallel respectively to a, β, γ.

Now if the vectors corresponding to those sides taken in order be aa, $b\beta$, $c\gamma$ respectively, we shall have, by going round the triangle,

$$aa + b\beta + c\gamma = 0.$$

10. If a, β, γ are three vectors neither parallel nor in the same plane, it is impossible to find numerical values of a, b, c, not equal to zero, which shall render $aa + b\beta + c\gamma = 0$.

For (Art. 5) $aa + b\beta$ can be represented by a third vector in the plane which contains two lines parallel respectively to a, β. Now $c\gamma$ is not in that plane, therefore (Art. 6) their sum cannot equal 0.

It follows that if $aa + b\beta + c\gamma = 0$ and a, β, γ are not parallel vectors, they are in the same plane.

11. There is but one way of making the sum of multiples of a, β, γ (as in Art. 9) equal to 0.

Let $aa + b\beta + c\gamma = 0$,

and also $pa + q\beta + r\gamma = 0$.

By eliminating γ we get

$$(ar - cp)\, a + (br - cq)\, \beta = 0 \; ;$$
$$\therefore \text{(Art. 6)} \;\; ar = cp, \;\; br = cq,$$
$$\text{or} \;\; a : b : c :: p : q : r,$$

so that the second equation is simply a multiple of the first.

12. If a, β, γ are coinitial, coplanar vectors terminating in a straight line, then the same values of a, b, c which render $aa + b\beta + c\gamma = 0$ will also render $a + b + c = 0$.

Let vector $OA = a$, $OB = \beta$, $OC = \gamma$, ABC being a straight line; then

$$AB = \beta - a,$$
$$AC = \gamma - a.$$

But AC is a multiple of AB,

or $\quad \gamma - a = p(\beta - a)$,

i.e. $\quad (p-1)a - p\beta + \gamma = 0$.

But $\qquad (p-1) - p + 1 = 0$;

and as $p-1$, $-p$, $+1$ correspond to a, b, c and satisfy the condition required, the proposition is proved generally (Art. 11).

13. Conversely, if a, β, γ are coinitial coplanar vectors, and if both $aa + b\beta + c\gamma = 0$ and $a + b + c = 0$, then do a, β, γ terminate in a straight line.

For $\qquad a\gamma + b\gamma + c\gamma = 0$;

therefore by subtraction

$$a(\gamma - a) + b(\gamma - \beta) = 0,$$

i.e. $\gamma - a$ is a multiple of $\gamma - \beta$, and therefore (Art. 4) in the same straight line with it: i.e. AC is in the same straight line with BC. (See Tait's *Quaternions*, § 30.)

EXAMPLES.

Ex. 1. *If two triangles are so situated that the lines which join corresponding angles meet in a point, then pairs of corresponding sides being produced will meet in a straight line.*

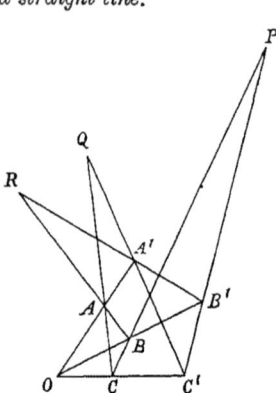

ABC, $A'B'C'$ are the triangles; O the point in which $A'A$, $B'B$, $C'C$ meet; P, Q, R the points in which BC, $B'C'$, &c. meet: PQR is a straight line.

Let $OA = a$, $OB = \beta$, $OC = \gamma$,

$\quad OA' = ma$, $OB' = n\beta$, $OC' = p\gamma$,

then $\qquad BA = a - \beta$,

and $\qquad BR = x(a - \beta)$;

$\qquad B'A' = ma - n\beta$,

and $\qquad B'R = y(ma - n\beta)$.

Now $BB' = BR - B'R$ gives

$$(n-1)\,\beta = x\,(a - \beta) - y\,(ma - n\beta)\,;$$

$$\therefore\ n - 1 = -x + ny,\ \ 0 = x - my,$$

and
$$x = -\frac{m\,(n-1)}{m-n}\,;$$

whence
$$OR = OB + BR = \beta - \frac{m\,(n-1)}{m-n}\,(a - \beta)$$

$$= \frac{n\,(m-1)\,\beta - m\,(n-1)\,a}{m-n}.$$

Similarly,
$$OP = \frac{p\,(n-1)\,\gamma - n\,(p-1)\,\beta}{n-p}\,,$$

$$OQ = \frac{m\,(p-1)\,a - p\,(m-1)\,\gamma}{p-m}\,;$$

$$\therefore\ (m-n)(p-1)\,OR + (n-p)(m-1)\,OP$$
$$+ (p-m)(n-1)\,OQ = 0.$$

And also

$$(m-n)(p-1) + (n-p)(m-1) + (p-m)(n-1) = 0,$$

whence (Art. 13) P, Q, R are in the same straight line.

Ex. 2. *If a quadrilateral be divided into two quadrilaterals by any cutting line, the centres of the three shall lie in a straight line.*

Let $P_1Q_1Q_3P_3$ be the quadrilateral divided into two by the

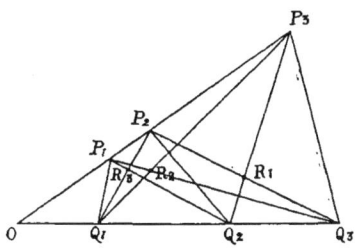

line P_2Q_2. Let the diagonals of $P_2Q_2Q_3P_3$ meet in R_1; and so of the others : R_1, R_2, R_3 are the centres.

Produce P_3P_1, Q_3Q_1 to meet in O. Let unit vectors along OP, OQ be denoted by a, β; and put

$$OP_1 = m_1 a, \quad OP_2 = m_2 a, \quad OP_3 = m_3 a;$$
$$OQ_1 = n_1 \beta, \quad OQ_2 = n_2 \beta, \quad OQ_3 = n_3 \beta;$$

then

$$OR_3 = OP_1 + P_1R_3 = m_1 a + x(n_2 \beta - m_1 a),$$

and

$$OR_3 = OQ_1 + Q_1R_3 = n_1 \beta + y(m_2 a - n_1 \beta).$$

Equating, we have

$$m_1 - m_1 x = m_2 y, \text{ and } n_2 x = n_1 - n_1 y;$$

$$\therefore \ x = \frac{(m_1 - m_2) n_1}{m_1 n_1 - m_2 n_2},$$

and

$$OR_3 = \frac{m_1 m_2 (n_1 - n_2) a + n_1 n_2 (m_1 - m_2) \beta}{m_1 n_1 - m_2 n_2}.$$

Similarly,

$$OR_1 = \frac{m_2 m_3 (n_2 - n_3) a + n_2 n_3 (m_2 - m_3) \beta}{m_2 n_2 - m_3 n_3},$$

$$OR_2 = \frac{m_1 m_3 (n_3 - n_1) a + n_1 n_3 (m_3 - m_1) \beta}{m_3 n_3 - m_1 n_1};$$

$$\therefore \ (m_1 n_1 - m_2 n_2) m_3 n_3 \, OR_3 + (m_2 n_2 - m_3 n_3) m_1 n_1 \, OR_1$$
$$+ (m_3 n_3 - m_1 n_1) m_2 n_2 \, OR_2 = 0.$$

And also

$$(m_1 n_1 - m_2 n_2) m_3 n_3 + (m_2 n_2 - m_3 n_3) m_1 n_1$$
$$+ (m_3 n_3 - m_1 n_1) m_2 n_2 = 0,$$

whence (Art. 13) R_1, R_2, R_3 are in the same straight line.

COR. R_1, R_2, R_3 will pass through O provided the coefficients of a and β in the three vectors have the same proportion, i.e. provided

$$\frac{1}{m_1} - \frac{1}{m_2} : \frac{1}{m_2} - \frac{1}{m_3} :: \frac{1}{n_1} - \frac{1}{n_2} : \frac{1}{n_2} - \frac{1}{n_3}.$$

Ex. 3. *If AD, BE, CF be drawn cutting one another at any point G within a triangle, then FD, DE, EF shall meet the third sides of the triangle produced in points which lie in a straight line.*

Also the produced sides of the triangle shall be cut harmonically.

If, as in Ex. 5, Art. 7, we put

$$DC = a, \quad CE = \beta, \quad BC = ma, \quad CA = n\beta,$$

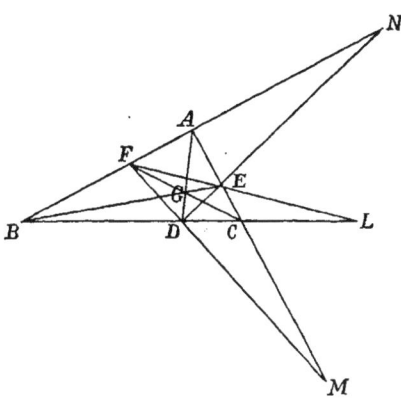

we get, as in that example,

$$AF \,:\, BF \,::\, n-1 \,:\, m-1\,;$$

$$\therefore BF = \frac{m-1}{m+n-2}\,(ma+n\beta),$$

and
$$FD = BD - BF = \frac{m-1}{m+n-2}\,\{(n-2)\,a - n\beta\}.$$

$DM = xFD,$ compared with

$$DM = DC - MC = a - y\beta,$$

gives
$$x\frac{(m-1)\,(n-2)}{m+n-2} = 1, \quad x\frac{(m-1)\,n}{m+n-2} = y\,;$$

$$\therefore y = \frac{n}{n-2},$$

and
$$BM = BC - MC = ma - \frac{n}{n-2}\,\beta.$$

Again,　$FE = FA + AE = \dfrac{n-1}{m+n-2}\{ma - (m-2)\,\beta\}.$

And $EL = xFE$, compared with

$$EL = CL - CE = ya - \beta,$$

gives

$$y = \frac{m}{m-2},$$

$$BL = (y + m) a = \frac{m(m-1)}{m-2} a.$$

Thirdly, $DN = xDE = x(a + \beta)$, compared with

$$DN = BN - BD = y(ma + n\beta) - (m-1) a,$$

gives

$$y = \frac{m-1}{m-n},$$

and

$$BN = \frac{m-1}{m-n} (ma + n\beta).$$

Now $(m-1)(n-2) BM + (m-n) BN$
$$- (m-2)(n-1) BL = 0.$$

Also $(m-1)(n-2) + (m-n) - (m-2)(n-1) = 0 ;$

therefore BM, BN, BL are in a straight line (Art. 13).

Further,

$$CL = \frac{m}{m-2} CD,$$

$$BL = \frac{m}{m-2} BD ;$$

$$\therefore CL : CD :: BL : BD,$$

and BL is cut harmonically.

Ex. 4. *The point of intersection of bisectors of the sides of a triangle from the opposite angles, the point of intersection of perpendiculars on the sides from the opposite angles, and the point of intersection of perpendiculars on the sides from their middle points, lie in a straight line which is trisected by the first of these points.*

1°. Let unit vector $CB = a$, unit vector $CA = \beta$,

then, Ex. 4, Art. 7, $CG = \dfrac{1}{3} (aa + b\beta)$.

2°. Let AH, BK perpendiculars on the sides intersect in O,

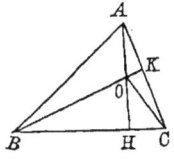

then
$$HA = b\beta - ba \cos C,$$
$$= b\,(\beta - a \cos C),$$
$$KB = a\,(a - \beta \cos C).$$

Now $CO = CA + AO$, and also $= CB + BO$, gives

$$b\beta + yb\,(\beta - aa \cos C) = aa + xa\,(a - \beta \cos C),$$

$$\therefore \; ax = \frac{b \cos C - a}{\sin^2 C},$$

and
$$CO = \frac{\cos C}{\sin^2 C}\{(b - a \cos C)\,a + (a - b \cos C)\,\beta\}.$$

3°. Let perpendiculars from D and E (Ex. 4, Art. 7) meet in X,

then DX is a multiple of HA.

$$\therefore \; CX = CD + DX = CE + EX \text{ gives}$$

$$\frac{1}{2}\,aa + v\,(\beta - a \cos C) = \frac{1}{2}\,b\beta + z\,(a - \beta \cos C),$$

$$\therefore \; v = \frac{b - a \cos C}{2 \sin^2 C},$$

and
$$CX = \frac{(a - b \cos C)\,a + (b - a \cos C)\,\beta}{2 \sin^2 C},$$

$$\therefore \; 2CX + CO - 3CG = 0,$$

and also
$$2 + 1 - 3 = 0,$$

$\therefore \; X$, O, G are in a straight line.

Also
$$CO - CG = 2\,(CG - CX),$$
$$\text{or vector } GO = 2 \text{ vector } XG,$$
$$\therefore \; GO = 2GX,$$

and G trisects XO.

14. The vector to the mean point of any polygon is the mean of the vectors to the angles of the polygon.

1^{0}. Let O be any point; then in the figure of Ex. 4, Art. 7 we have, calling OA, a, OB, β and OC, γ,

$$OG = a + AG = \beta + BG = \gamma + CG$$

$$= \frac{1}{3}(a + \beta + \gamma) + \frac{1}{3}(AG + BG + CG)$$

$$= \frac{1}{3}(a + \beta + \gamma);$$

because
$$AG + BG + CG = \frac{2}{3}(AD + BE + CF)$$

$$= \frac{2}{3}\{(AB + AC) + (BA + BC) + (CA + CB)\}$$

$$= 0.$$

2^{0}. If OA, OB, OC, OD be a, β, γ, δ, in the figure of Ex. 7, Art. 7, we have

$$OX = OH + HX = OH + \frac{1}{2}(OF - OH)$$

$$= \frac{1}{2}(OF + OH) = \frac{1}{4}(a + \beta + \gamma + \delta).$$

3^{0}. In the more general case we may define the mean point in a manner analogous to that adopted in mechanics to define the centre of inertia of equal masses placed at the angular points of the figure. Thus, if we take any rectangular axes OX, OY, and designate by a, β unit vectors parallel to these axes; and by ρ_1, ρ_2, &c. the vectors to the different points; and if we write x_1, y_1; x_2, y_2, &c. for the Cartesian co-ordinates of the different points referred to those axes; and define the mean point as the centre of inertia of equal masses placed at the angular points; the Cartesian co-ordinates of that point will be

$$x = \frac{x_1 + x_2 + \ldots}{m}, \quad y = \frac{y_1 + y_2 + \ldots}{m},$$

and its vector
$$\rho = xa + y\beta.$$

Now $\qquad \rho_1 = x_1 \alpha + y_1 \beta, \;\; \rho_2 = x_2 \alpha + y_2 \beta, \; \&\text{c.}$

$$\therefore \frac{\rho_1 + \rho_2 + \dots}{m} = \frac{x_1 + x_2 + \dots}{m}\, \alpha + \frac{y_1 + y_2 + \dots}{m}\, \beta$$

$$= x\alpha + y\beta,$$

$$= \rho.$$

COR. 1. $\qquad (\rho_1 - \rho) + (\rho_2 - \rho) + (\rho_3 - \rho) + \&\text{c.} = 0,$

i.e. the sum of the vectors of all the points, drawn from the mean point, $= 0$.

The extension of the same theorem to three dimensions is obvious.

COR. 2. If we have another system of n points whose vectors are σ_1, σ_2, &c. then the vector to the mean point is

$$\sigma = \frac{\sigma_1 + \sigma_2 + \dots}{n}.$$

If now τ be the mean point of the whole system, we have

$$\tau = \frac{\rho_1 + \rho_2 + \dots + \sigma_1 + \sigma_2 + \dots}{m + n},$$

or $\qquad (m + n)\,\tau - m\rho - n\sigma = 0,$

hence (13) τ, ρ, σ terminate in a right line; or the general mean point is situated on the right line which connects the two partial mean points.

ADDITIONAL EXAMPLES TO CHAP. II.

1. If P, Q, R, S be points taken in the sides AB, BC, CD, DA of a parallelogram, so that $AP : AB :: BQ : BC$, &c., $PQRS$ will form a parallelogram.

2. If the points be taken so that $AP = CR$, $BQ = DS$, the same is true.

3. The mean point of $PQRS$ is in both cases the same as that of $ABCD$.

4. If $P'Q'R'S'$ be another parallelogram described as in Ex. 1, the intersections of PQ, $P'Q'$, &c. shall be in the angular points of a parallelogram $EFGH$ constructed from $PQRS$ as $P'Q'R'S'$ is constructed from $ABCD$.

5. The quadrilateral formed by bisecting the sides of a quadrilateral and joining the successive points of bisection is a parallelogram, with the same mean point.

6. If the same be true of any other equable division such as trisection, the original quadrilateral is a parallelogram.

⋄ 7. If any line pass through the mean point of a number of points, the sum of the perpendiculars on this line from the different points, measured in the same direction, is zero.

8. From a point E in the common base AB of the two triangles ABC, ABD, straight lines are drawn parallel to AC, AD, meeting BC, BD at F, G; shew that FG is parallel to CD.

9. From any point in the base of a triangle, straight lines are drawn parallel to the sides: shew that the intersections of the diagonals of every parallelogram so formed lie in a straight line.

⋅ 10. If the sides of a triangle be produced, the bisectors of the external angles meet the opposite sides in three points which lie in a straight line.

• 11. If straight lines bisect the interior and exterior angles at A of the triangle ABC in D and E respectively; prove that BD, BC, BE form an harmonical progression.

12. The diagonals of a parallelepiped bisect one another.

13. The mean point of a tetrahedron is the mean point of the tetrahedron formed by joining the mean points of the triangular faces; and also those of the edges.

14. If the figure of Ex. 11, Art. 7, be that of a gauche quadrilateral (a term employed by Chasles to signify that the triangles

AOD, BOD are not in the same plane), the lines QP, DO, RS will meet in a point, provided

$$\frac{AP}{PD} = m\,\frac{OS}{SB}, \text{ and } \frac{AQ}{QO} = m\,\frac{DR}{RB}.$$

15. If through any point within the triangle ABC, three straight lines MN, PQ, RS be drawn respectively parallel to the sides AB, AC, BC; then will

$$\frac{MN}{AB} + \frac{PQ}{AC} + \frac{RS}{BC} = 2.$$

16. $ABCD$ is a parallelogram; E, the point of bisection of AB; prove that AC, DE being joined will trisect each other.

17. $ABCD$ is a parallelogram; PQ any line parallel to CD; PD, QC meet in S, PA, QB in R; prove that AD is parallel to RS.

CHAPTER III.

VECTOR MULTIPLICATION AND DIVISION.

15. WE trust we have made the reader understand by what we stated in our Introductory Chapter, that, whilst we retain for 'multiplication' all its old properties, so far as it relates to ordinary algebraical quantities, we are at liberty to attach to it any signification we please when we speak of the multiplication of a vector by or into another vector. Of course the interpretation of our results will depend on the definition, and may in some points differ from the interpretation of the results of multiplication of numerical quantities.

It is necessary to start with one limitation. Whereas in Algebra we are accustomed to use at random the phrases 'multiply by' and 'multiply into' as tantamount to the same thing, it is now impossible to do so. We must select one to the exclusion of the other. The phrase selected is 'multiply into'; thus we shall understand that the first written symbol in a sequence is the operator on that which follows: in other words that $\alpha\beta$ shall read 'α into β', and denote α operating on β.

16. As in the Cartesian Geometry, so here we indicate the position of a point in space by its relation to three axes, mutually at right angles, which we designate the axes of x, y, and z respectively. For graphic representation the axes of x and y are drawn in the plane of the paper whilst that of z being perpendicular to that plane is drawn in perspective only. As in ordinary

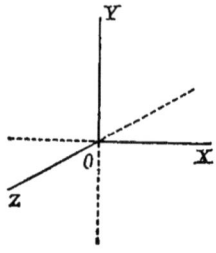

geometry we assume that when vectors measured forwards are represented by positive symbols, vectors measured backwards will be represented by the corresponding negative symbols. In the figure before us, the positive directions are *forwards, upwards* and *outwards;* the corresponding negative directions, *backwards, downwards* and *inwards.*

With respect to vector rotation we assume that, looked at in perspective in the figure before us, it is negative when in the direction of the motion of the hands of a watch, positive when in the contrary direction. In other words, we assume, as is done in modern works on Dynamics, that rotation is positive when it takes place from y to z, z to x, x to y : negative when it takes place in the contrary directions (see *Tait*, Art. 65).

Unit vectors at right angles to each other.

17. Definition. If i, j, k be unit vectors along Ox, Oy, Oz respectively, the result of the multiplication of i into j or ij is defined to be the turning of j through a right angle in the plane perpendicular to i and in the positive direction; in other words, the operation of i on j turns it round so as to make it coincide with k; and therefore briefly $ij = k$.

To be consistent it is requisite to admit that if i instead of operating on j had operated on any other unit vector perpendicular to i in the plane of yz, it would have turned it through a right angle in the same direction, so that ik can be nothing else than $-j$. Extending to other unit vectors the definition which we have illustrated by referring to i, it is evident that j operating on k must bring it round to i, or $jk = i$.

Again, always remembering that the positive directions of rotation are y to z, z to x, x to y, we must have $ki = j$.

18. As we have stated, we retain in connection with this definition the old laws of numerical multiplication, whenever numerical quantities are mixed up with vector operations; thus $2i . 3j = 6ij$. Further, there can be no reason whatever, but the contrary, why the laws of addition and subtraction should undergo

any modification when the operations are subject to this new
definition; we must clearly have

$$i\,(j + k) = ij + ik.$$

Finally, as we are to regard the operations of this new definition as operations of multiplication—magnitude and motion
of rotation being united in one vector symbol as multiplier,
just as magnitude and motion of translation were united in
one vector symbol in the last chapter—we are bound to retain
all the laws of algebraic multiplication so far as they do not
give results inconsistent with each other. In no other way can
the conclusions be made to compare with those deduced from
the corresponding operations in the previous science. Thus we
retain what Sir William Hamilton terms the *associative law of
multiplication*: the law which assumes that it is indifferent in
what way operations are grouped, provided the order be not
changed; the law which makes it indifferent whether we consider
abc to be $a \times bc$ or $ab \times c$. This law is *assumed* to be applicable to
multiplication in its new aspect (for example that $ijk = ij \cdot k$), and
being assumed it limits the science to certain boundaries, and,
along with other assumed laws, furnishes the key to the interpretation of results.

The law is by no means a necessary law. Some new forms of
the science may possibly modify it hereafter. In the meantime
the assumption of the law fixes the limits of the science.

The *commutative* law of multiplication under which order may
be deranged, which is assumed as the groundwork of common
algebra (we say *assumed* advisedly) is now no longer tenable. And
this being the case it is found that the science of Quaternions
breaks down one of the barriers imposed by this law and expands
itself into a new field.

ij is *not* equal to ji, it is clearly impossible it should be.

A simple inspection of the figure, and a moment's consideration
of the definition, will make this plain. The definition imposes on i
as an operator on j the duty of turning j through a right angle as
if by a left-handed turn with a cork-screw handle, thus throwing
j up from the plane xy; when, on the other hand, j is the operator

and i the vector operated on, a similar left-handed turn will bring i *down* from the plane of xy. In fact $ij = k$, $ji = -k$, and so $ij = -ji$.

19. We go on to obtain one or two results of the application of the associative law.

1. Since $ij = k$, we have $i \cdot ij = ik = -j$.

Now by the law in question,

$$i \cdot ij = ii \cdot j = i^2 \cdot j ;$$
$$\therefore \ i^2 \cdot j = -j,$$

or
$$i^2 = -1.$$

Our first result is that the square of the unit vector along Ox is -1; and as Ox may have any direction whatever, we have, generally, *the square of a unit vector* $= -1$. In other words, the repetition of the operation of turning through a right angle reverses a vector.

2. Again, $ijk = i \cdot jk = i \cdot i = i^2 = -1.$

Similarly it may be proved that

$$jki = kij = -1,$$

or no change is produced in the product so long as direct cyclical order is maintained.

3. But $ikj = i \cdot kj = i \cdot -i = -i^2 = +1 ;$
$$\therefore \ ijk = -ikj,$$

or a derangement of cyclical order changes the sign of the product.

This last conclusion is also manifest from Art. 18.

Vectors generally not at right angles to each other.

20. We have already (Art. 8) laid down the principle of separation of the vector into the product of tensor and unit vector; and we apply this to multiplication by the considerations given in Art. 18, from which it follows at once that if a be a vector along Ox containing a units, β a vector along Oy containing b units,

$$a = ai, \ \beta = bj, \text{ and } a\beta = abij.$$

In the same way
$$a^2 = ai \cdot ai = a^2 i^2 = -a^2,$$
or the square of a *vector* is the square of the corresponding *line* with the negative sign.

Seeing therefore the facility with which we can introduce tensors whenever wanted, we may direct our principal attention, as far as multiplication is concerned, to unit vectors.

21. We proceed then next to find the product $\alpha\beta$, when α and β are vectors not at right angles to one another.

1.   Let α, β be unit vectors.

Let $OA = \alpha$, $OB = \beta$.

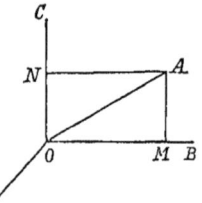

Take $OC = \gamma$, a unit vector perpendicular to OB and in the plane BOA. Take also DO or DO produced $= \epsilon$, a unit vector perpendicular to the plane BOA.

Draw AM, AN perpendicular to OB, OC, and let the angle $BOA = \theta$; then

$$\text{vector } OA = OM + MA = OM + ON \text{ (Art. 1)}$$
$$= \text{part of } OB + \text{part of } OC \text{ (Art. 3).}$$

Now it is evident that OM as a line is that part of OB which is represented by the multiplier $\cos\theta$, or $OM = OB\cos\theta$, and similarly that $ON = OC\sin\theta$: consequently (Art. 3) the same applies to them as vectors; i.e.

$$\text{vector } OM = \beta\cos\theta, \text{ vector } ON = \gamma\sin\theta;$$
$$\therefore \alpha = \beta\cos\theta + \gamma\sin\theta,$$
and
$$\alpha\beta = (\beta\cos\theta + \gamma\sin\theta)\beta$$
$$= \beta^2\cos\theta + \gamma\beta\sin\theta.$$
But
$$\beta^2 = -1 \text{ (19. 1),}$$
$$\gamma\beta = \epsilon \text{ (17);}$$

[Observe that γ, β and ϵ of the present Article correspond to j, i and $-k$ of Art. 17.]

$$\therefore \alpha\beta = -\cos\theta + \epsilon\sin\theta.$$

2.   If a, β are not unit vectors, but contain Ta and $T\beta$ units respectively, we have at once, by the principle laid down in Art. 20,

$$a\beta = TaT\beta \, (- \cos \theta + \epsilon \sin \theta).$$

3.   It thus appears that the product of two vectors a, β not at right angles to each other consists of two distinct parts, a numerical quantity and a vector perpendicular to the plane of a, β. The former of these Sir William Hamilton terms the SCALAR part, the latter the VECTOR part.   We may now write

$$a\beta = Sa\beta + Va\beta,$$

where S is read scalar, V vector : and we find

$$Sa\beta = - \, TaT\beta \cos \theta,$$

$$Va\beta = TaT\beta \, \epsilon \sin \theta.$$

4.   The coefficient of ϵ in $Va\beta$ is the area of the parallelogram whose sides are equal and parallel to the lines of which a, β are the vectors.

22.   To obtain βa we have, a and β being unit vectors,

$$a = \beta \cos \theta + \gamma \sin \theta \; ;$$

$$\therefore \; \beta a = \beta \, (\beta \cos \theta + \gamma \sin \theta)$$

$$= \beta^2 \cos \theta + \beta\gamma \sin \theta$$

$$= - \cos \theta - \epsilon \sin \theta \, (\text{Art. 19. 1 and 18}) \, ;$$

therefore generally

$$\beta a = TaT\beta \, (- \cos \theta - \epsilon \sin \theta).$$

It is scarcely necessary to remark that whilst γ operating on β turns it inwards from OB to DO produced, β operating on γ turns it outwards from OC to OD, causing it to become $- \epsilon$.

We have therefore

1. $Sa\beta = S\beta a$.

2. $Va\beta = - V\beta a$.

3. $a\beta + \beta a = 2Sa\beta$.

4. $a\beta - \beta a = 2Va\beta$.

5. $(a + \beta)^2 = (a + \beta)(a + \beta)$
$$= a^2 + a\beta + \beta a + \beta^2$$
$$= a^2 + 2Sa\beta + \beta^2.$$

6. $(a - \beta)^2 = a^2 - 2Sa\beta + \beta^2.$

7.   If a, β are at right angles to each other, $Sa\beta = 0$, and conversely.

8. $Va\beta$ is a vector in the direction perpendicular to the plane which passes through a, β.

9. $a^2\beta^2 = a\beta \cdot \beta a$ because β^2 is a scalar;
$$\therefore \ a^2\beta^2 = (Sa\beta + Va\beta)(Sa\beta - Va\beta)$$
$$= (Sa\beta)^2 - (Va\beta)^2.$$

Note. $a^2\beta^2$ must not be confounded with $(a\beta)^2$.

23.   Before proceeding further it is desirable we should work out a few simple Examples.

Ex. 1.   *To express the cosine of an angle of a triangle in terms of the sides.*

Let ABC be a triangle; and retaining the usual notation of Trigonometry, let
$$CB = a, \quad CA = \beta;$$
then $(\text{vector } AB)^2 = (a - \beta)^2$
$$= a^2 - 2Sa\beta + \beta^2 \ (22. \ 6),$$
or, changing all the signs to pass from vectors to lines (20) and applying 21. 3,
$$c^2 = a^2 - 2ab \cos C + b^2.$$

Ex. 2.   *To express the relations between the sides and opposite angles of a triangle.*

Let $CB = a, \quad CA = \beta, \quad BA = \gamma.$
Then $CB + BA = CA$ gives
$$a + \gamma = \beta,$$
$$a = \beta - \gamma;$$
$$\therefore \ a^2 = a(\beta - \gamma) = a\beta - a\gamma.$$
Take the vectors of each side.

Now $V\alpha^2 \neq 0$, for $\alpha^2 = -a^2$ has no vector part,

$$\therefore \quad V\alpha\beta = V\alpha\gamma \; ;$$

i. e. (21. 3) $ab\epsilon \sin C = ac\epsilon \sin B,$

or $b \sin C = c \sin B \; ;$

i.e. $b : c :: \sin B : \sin C.$

Ex. 3. *The sum of the squares of the diagonals of a parallelogram is equal to the sum of the squares of the sides.*

Retaining the notation and figure of Ex. 1, Art. 7,

$$CB = \alpha + \beta,$$
$$AD = \alpha - \beta \; ;$$
$$\therefore \quad CB^2 + DA^2 = 2\alpha^2 + 2\beta^2,$$

and, changing all the signs, we get (20) for the corresponding lines,

$$CB^2 + DA^2 = 2CA^2 + 2AB^2$$
$$= CA^2 + AB^2 + BD^2 + DC^2.$$

Ex. 4. *Parallelograms upon the same base and between the same parallels are equal.*

It is necessary to remind the reader of what we have already stated, that examples such as this are given for illustration only. We *assume* that the area of the parallelogram is the product of two adjacent sides and the sine of the contained angle.

Adopting the figure of Euclid I. 35 and writing $TV\beta\alpha$ as the tensor multiplier of $V\beta\alpha$ so as to drop the vector ϵ on both sides; we have, calling BA, α; BC, β;

$$BE = BA + AE$$
$$= \alpha + x\beta \; ;$$
$$\therefore \quad V . \beta (\alpha + x\beta) = V (BC . BE),$$
$$\text{i.e.} \quad V\beta\alpha = V(BC. BE),$$

remembering that $x\beta^2$ has no vector part.

Hence $\quad T . V\beta\alpha = T (BC . BE),$

i.e. $BC . BA \sin ABC = BC . BE \sin EBC$ (21. 3),

which proves the proposition.

Ex. 5. *On the sides AB, AC of a triangle are constructed any two parallelograms ABDE, ACFG : the sides DE, FG are produced to meet in H. Prove that the sum of the areas of the parallelograms ABDE, ACFG is equal to the area of the parallelogram whose adjacent sides are respectively equal and parallel to BC and AH.*

Let $\qquad BA = a, \quad AE = \beta, \quad AC = \gamma, \quad GA = \delta,$

then $\qquad AH = \beta + xa, \text{ and } AH = -\delta - y\gamma ;$

$$\therefore \ VaAH = Va\beta \text{ and } V\gamma AH = -V\gamma\delta$$
$$= V\delta\gamma \ (22. \ 2),$$

hence $\qquad V(a + \gamma) AH = Va\beta + V\delta\gamma,$

i.e. (21. 4), the parallelogram whose sides are parallel and equal to BC, AH, equals the two parallelograms whose sides are parallel and equal to BA, AE ; GA, AC respectively.

[The reader is requested to notice that the *order GA, AC* is the same as the order *BA, AE*, and *BA, AH* : so that the vector ϵ is common to all.]

Ex. 6. *If O be any point whatever either in the plane of the triangle ABC or out of that plane, the squares of the sides of the triangle fall short of three times the squares of the distances of the angular points from O, by the square of three times the distance of the mean point from O.*

Let $\qquad OA = a, \quad OB = \beta, \quad OC = \gamma,$

then (Art. 14), $\qquad OG = \dfrac{1}{3}(a + \beta + \gamma),$

or $\qquad a^2 + \beta^2 + \gamma^2 + 2S(a\beta + \beta\gamma + \gamma a) = 9OG^2.$

Now $\qquad AB = \beta - a, \quad BC = \gamma - \beta, \quad CA = a - \gamma,$

$$\therefore \ AB^2 + BC^2 + CA^2 = 2(a^2 + \beta^2 + \gamma^2) - 2S(a\beta + \beta\gamma + \gamma a)$$
$$= 3(a^2 + \beta^2 + \gamma^2) - 9OG^2,$$

and the lines

$$AB^2 + BC^2 + CA^2 = 3(OA^2 + OB^2 + OC^2) - (3OG)^2.$$

Ex. 7. *The sum of the squares of the distances of any point O from the angular points of the triangle exceeds the sum of the squares of its distances from the middle points of the sides by the sum of the squares of half the sides.*

Retaining the notation of the last example, and the figure of Ex. 4, Art. 7,

$$OD = \frac{1}{2}(\beta + \gamma), \;\; OE = \frac{1}{2}(\gamma + \alpha), \;\; OF = \frac{1}{2}(\alpha + \beta);$$

$$\therefore \; 4(OD^2 + OE^2 + OF^2) = 2(\alpha^2 + \beta^2 + \gamma^2) + 2S(\alpha\beta + \beta\gamma + \gamma\alpha)$$
$$= \alpha^2 + \beta^2 + \gamma^2 + 9OG^2$$
$$= 4(\alpha^2 + \beta^2 + \gamma^2) - (AB^2 + BC^2 + CA^2);$$

$$\therefore \text{ as lines } OD^2 + OE^2 + OF^2 + \frac{AB^2 + BC^2 + CA^2}{4} = OA^2 + OB^2 + OC^2.$$

Ex. 8.   *The squares of the sides of any quadrilateral exceed the squares of the diagonals by four times the square of the line which joins the middle points of the diagonals.*

Retaining the figure and notation of Ex. 8, Art. 7, we have squares of sides as vectors

$$= \alpha^2 + (\beta - \alpha)^2 + (\gamma - \beta)^2 + \gamma^2$$
$$= 2(\alpha^2 + \beta^2 + \gamma^2) - 2S(\alpha\beta + \beta\gamma),$$

and squares of diagonals

$$= \beta^2 + (\gamma - \alpha)^2$$
$$= \alpha^2 + \beta^2 + \gamma^2 - 2S\alpha\gamma;$$

therefore the former sum exceeds the latter by

$$\alpha^2 + \beta^2 + \gamma^2 - 2S\alpha\beta - 2S\beta\gamma + 2S\alpha\gamma$$
$$= (\alpha + \gamma - \beta)^2$$
$$= 4\left(\frac{\alpha + \gamma}{2} - \frac{\beta}{2}\right)^2$$
$$= 4(OQ - OP)^2$$
$$= 4PQ^2.$$

Therefore as lines the same is true.

Note.   The points A, B, C, D need not be in one plane.

Ex. 9. *Four times the squares of the distances of any point whatever from the angular points of a quadrilateral are equal to the sum of the squares of the sides, the squares of the diagonals and the square of four times the distance of the point from the mean point of the figure.*

With the notation of Art. 14, and the figure of Ex. 7, Art. 7, we have

squares of the sides + squares of the diagonals

$$= (\beta - a)^2 + (\gamma - \beta)^2 + (\delta - \gamma)^2 + (a - \delta)^2 + (\gamma - a)^2 + (\delta - \beta)$$

$$= 3(a^2 + \beta^2 + \gamma^2 + \delta^2) - 2S(a\beta + a\gamma + a\delta + \beta\gamma + \beta\delta + \gamma\delta).$$

Now (Art. 14) $(a + \beta + \gamma + \delta)^2 = (4OX)^2;$

$$\therefore \ (4OX)^2 + \text{squares of sides} + \text{squares of diagonals}$$

$$= 4(OA^2 + OB^2 + OC^2 + OD^2).$$

Ex. 10. *The lines which join the mean points of three equilateral triangles described outwards on the three sides of any triangle form an equilateral triangle whose mean point is the same as that of the given triangle.*

Let P, Q, R be the mean points of the equilateral triangles on BC, CA, AB; $PD = a$, $DC = \beta$, $CE = \gamma$, $EQ = \delta$; and let the sides of the triangle ABC be $2a$, $2b$, $2c$.

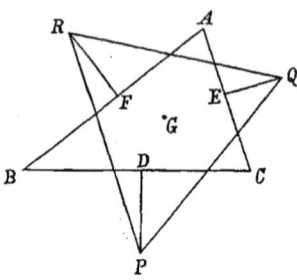

$$\therefore \ PQ^2 = (a + \beta + \gamma + \delta)^2$$

$$= a^2 + \beta^2 + \gamma^2 + \delta^2 + 2Sa\beta + 2Sa\gamma + 2Sa\delta$$

$$+ 2S\beta\gamma + 2S\beta\delta + 2S\gamma\delta.$$

Changing all the signs and observing that

$$S\alpha\beta = 0, \quad Sa\gamma = -\frac{2}{\sqrt{3}}\, ab \sin C, \&c.$$

we have (writing the results in the same order),

$$\text{line } PQ^2 = \frac{a^2}{3} + a^2 + b^2 + \frac{b^2}{3} + 0$$

$$+ \frac{2}{\sqrt{3}}\, ab \sin C + \frac{2}{3}\, ab \cos C - 2ab \cos C + \frac{2}{\sqrt{3}}\, ab \sin C + 0$$

$$= \frac{4}{3}\,(a^2 + b^2 - ab \cos C) + \frac{4}{\sqrt{3}}\, ab \sin C$$

$$= \frac{2}{3}\,(a^2 + b^2 + c^2) + \frac{2}{\sqrt{3}}\, \text{area of } ABC,$$

which being symmetrical in a, b, c proves that PQR is equilateral.

Again, G being the mean point of ABC,

$$PG = PD + DG = \alpha + \frac{\beta}{3} + \frac{2\gamma}{3},$$

$$\therefore\ PG^2 = a^2 + \frac{\beta^2}{9} + \frac{4\gamma^2}{9} + \frac{2}{3}\,S\alpha\beta + \frac{4}{3}\,Sa\gamma + \frac{4}{9}\,S\beta\gamma,$$

and line $PG^2 = \dfrac{a^2}{3} + \dfrac{a^2}{9} + \dfrac{4b^2}{9} + \dfrac{4}{3\sqrt{3}}\, ab \sin C - \dfrac{4}{9}\, ab \cos C$

$$= \frac{2}{9}\,(a^2 + b^2 + c^2) + \frac{2}{3\sqrt{3}}\, \text{area } ABC;$$

$$\therefore\ PG = QG = RG;$$

and G is the mean point of the equilateral triangle PQR.

Ex. 11.   *In any quadrilateral prism, the sum of the squares of the edges exceeds the sum of the squares of the diagonals by eight times the square of the straight line which joins the points of intersection of the two pairs of diagonals.*

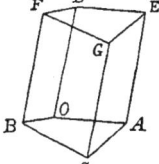

Let $OA = a$, $OB = \beta$, $OC = \gamma$, $OD = \delta$; sum of squares of edges =

$$2\left\{a^2 + \beta^2 + (\gamma - a)^2 + (\gamma - \beta)^2 + 2\delta^2\right\}$$

$$= 2\left\{2a^2 + 2\beta^2 + 2\gamma^2 + 2\delta^2 - 2Sa\gamma - 2S\beta\gamma\right\},$$

sum of squares of diagonals

$$= (\delta + \gamma)^2 + (\delta - \gamma)^2 + (\delta + a - \beta)^2 + (\delta + \beta - a)^2$$
$$= 2 \{a^2 + \beta^2 + \gamma^2 + 2\delta^2 - 2Sa\beta\}.$$

Also
$$\frac{1}{2} OG = \frac{1}{2} (\delta + \gamma)$$

$$= \text{vector to the point of bisection of}$$

CD, and therefore to the point of intersection of OG, CD, and vector from O to the point of bisection of AF, as also to that of BE, and therefore to the intersection of AF, BE

$$= \frac{1}{2} (\delta + a + \beta),$$

hence vector which joins the points of intersection of diagonals

$$= \frac{1}{2} (a + \beta - \gamma),$$

eight times the square of this vector

$$= 2 (a^2 + \beta^2 + \gamma^2 + 2Sa\beta - 2Sa\gamma - 2S\beta\gamma),$$

which, added to the sum of the squares of the diagonals, makes up the sum of the squares of the edges.

24. DEFINITION. We define the quotient or fraction $\dfrac{\beta}{a}$, where a and β are unit vectors, to be such that when it operates *on a* it produces β or $\dfrac{\beta}{a} . a = \beta$. This form of the definition enables us to strike out a by a dash made in the direction of ordinary writing, thus $\dfrac{\beta}{a} . a = \beta$. $\dfrac{\beta}{a}$ is therefore that multiplier which, operating on a, or on $\beta \cos \theta + \gamma \sin \theta$ (21), produces β.

Now $\cos \theta + \epsilon \sin \theta$ operating on $\beta \cos \theta + \gamma \sin \theta$ produces

$$\beta \cos^2 \theta + (\gamma + \epsilon\beta) \sin \theta \cos \theta + \epsilon\gamma \sin^2 \theta.$$

But a glance at the figure (Art. 21) will shew that

$$\epsilon\beta = -\gamma,$$

and
$$\epsilon\gamma = \beta ;$$

$\therefore \cos\theta + \epsilon\sin\theta$ operating on $\beta\cos\theta + \gamma\sin\theta$ produces β ;

hence $$\frac{\beta}{\alpha} = \cos\theta + \epsilon\sin\theta.$$

It may be worth while to exhibit another demonstration of this proposition : thus

$$\frac{\beta}{\alpha} \cdot \alpha\beta = \beta \cdot \beta \text{ (by the associative law)} = -1 \cdot (19 \cdot 1).$$

$$\text{i.e. } (21 \cdot 1) \frac{\beta}{\alpha} \cdot (-\cos\theta + \epsilon\sin\theta) = -1.$$

Now $$(\cos\theta + \epsilon\sin\theta)(-\cos\theta + \epsilon\sin\theta)$$
$$= -\cos^2\theta - \sin^2\theta$$
$$= -1;$$
$$\therefore \frac{\beta}{\alpha} = \cos\theta + \epsilon\sin\theta.$$

Cor. $\dfrac{\beta}{\alpha} = -\beta\alpha$ (by 22).

25. 1. DEFINITION. Still retaining α, β as unit vectors, since $\dfrac{\beta}{\alpha}$ operating on α causes it to become β, it may be defined as a VERSOR acting as if its axis were along OD (Fig. Art. 21). By comparing the result of that article with the definitions of Art. 17, it is clear that $\dfrac{\beta}{\alpha}$ or $\cos\theta + \epsilon\sin\theta$ is an operator of the same character as $-k$ or ϵ (as we have now called the corresponding unit vector) ; with this difference only, that whereas $-k$ or ϵ as an operator would turn α through a right angle, $\cos\theta + \epsilon\sin\theta$ turns it, in the same direction, only through the angle θ : $\cos\theta + \epsilon\sin\theta$ is then the *versor* through the angle θ.

2. If α, β are not unit vectors, the considerations already advanced render it evident that

$$\frac{\beta}{\alpha} = \frac{T\beta}{T\alpha}(\cos\theta + \epsilon\sin\theta).$$

Now $\dfrac{T\beta}{T\alpha}$ is itself of the nature of a tensor, for it is a numerical quantity, hence $\dfrac{\beta}{\alpha}$ is the product of a tensor and a versor.

26. By comparing the last Article with Art. 22 it appears that generally the product or quotient of two vectors may be expressed as the product of a tensor and a versor. This product Sir W. Hamilton names a QUATERNION.

COR. It is evident that a quaternion is also the sum of a scalar and a vector.

27. (1) If a, β, γ are unit vectors in the same plane, ϵ a unit vector perpendicular to that plane; we have seen that $\dfrac{\beta}{a}$ operating on a turns it round about ϵ as an axis to bring it into the position β. If now $\dfrac{\gamma}{\beta}$ be a second operator about the same axis in the same direction acting on β, it will bring it into the position γ. But it is evident that $\dfrac{\gamma}{a}$ acting on a would at once have brought it into the position γ. This is equivalent to the fact that $\dfrac{\gamma}{\beta} \cdot \dfrac{\beta}{a} = \dfrac{\gamma}{a}$; or in another form (Art. 24) that

$$(\cos\phi + \epsilon\sin\phi)(\cos\theta + \epsilon\sin\theta) = \cos(\theta+\phi) + \epsilon\sin(\theta+\phi).$$

From this it is evident that the results of Demoivre's Theorem apply to the form $\cos\theta + \epsilon\sin\theta$.

Further, it is evident that since $\cos\theta + \epsilon\sin\theta$ operating with ϵ as its axis, turns a vector through the angle θ, whilst ϵ itself acting in the same direction turns it through a right angle, $\cos\theta + \epsilon\sin\theta$ is *part* of the operation designated by ϵ, viz. that part which bears to the whole the proportion that θ bears to a right angle.

(2) Remembering then that the operations are of the nature of multiplication, it becomes evident that $\cos\theta + \epsilon\sin\theta$ as an operator may be abbreviated by $\epsilon^{\frac{\theta}{\frac{\pi}{2}}}$ or $\epsilon^{\frac{2\theta}{\pi}}$.

And since

$$(\cos\theta + \epsilon\sin\theta)(\cos\phi + \epsilon\sin\phi) = \cos(\theta+\phi) + \epsilon\sin(\theta+\phi),$$

we shall have

$$\epsilon^{\frac{2\theta}{\pi}} \cdot \epsilon^{\frac{2\phi}{\pi}} = \epsilon^{\frac{2}{\pi}(\theta+\phi)},$$

or the *law of indices* is applicable to this operator.

(3)   Now we have already seen (19. 1) that $\epsilon^2 = -1$;

$$\therefore \ \epsilon^4 = +1.$$

Conversely, if $\epsilon^n = \pm \epsilon$, n must be an odd number ; if $\epsilon^n = -1$, n must be an odd multiple of 2 ; and if $\epsilon^n = +1$, n must be an even multiple of 2.

(4)   When a, β are not units, the introduction of the corresponding tensor can be at once effected.

We conclude that a quaternion may be expressed as the *power* of a vector, to which the algebraic definition of an index is applicable.

28.   *Reciprocals* of quaternions—unit vectors.

1.   Since $\qquad\qquad a \cdot a = a^2 = -1,$

and $\qquad\qquad \dfrac{1}{a} \cdot a = 1$ (Def. Art. 24)

$$= -a \cdot a ;$$

$$\therefore \ \frac{1}{a} = -a, \text{ or } a^{-1} = -a ;$$

or the reciprocal of a unit vector is a unit vector in the opposite direction.

2.   Again, $\qquad a \cdot \dfrac{1}{a} = a\,(-a) = 1 = \dfrac{1}{a} \cdot a ;$

or a vector is commutative with its reciprocal.

3.   If q be a versor $\left(\text{say } \cos\theta + \epsilon\sin\theta, \text{ or } \dfrac{\beta}{a}\right),$

$$\frac{1}{q} \cdot q = 1 \ \text{(Def. extended)}.$$

Now $\qquad\qquad \dfrac{\beta}{a} = q ;$

$$\therefore \ \beta = qa, \text{ by operating on } a.$$

Also
$$\frac{a}{\beta} = \frac{1}{q},$$

$$a = \frac{1}{q}\beta, \text{ by operating on } \beta,$$

and
$$\beta = qa = q \cdot \frac{1}{q}\beta;$$

$$\therefore q \cdot \frac{1}{q} = 1 = \frac{1}{q} \cdot q,$$

or q and $\frac{1}{q}$ are commutative.

This is perhaps better demonstrated by observing that

$$\frac{\beta}{a} \cdot \frac{a}{\beta} = \frac{\beta}{\beta} = 1;$$

or that if
$$\frac{\beta}{a} = \cos\theta + \epsilon \sin\theta,$$

then must
$$\frac{a}{\beta} = \cos\theta - \epsilon \sin\theta;$$

factors which are from their very nature commutative.

As a verification, we have

$$\frac{\beta}{a} \cdot \frac{a}{\beta} = (\cos\theta + \epsilon \sin\theta)(\cos\theta - \epsilon \sin\theta)$$

$$= (\cos\theta)^2 - \epsilon^2 (\sin\theta)^2$$

$$= 1$$

because $\epsilon^2 = -1$ (28. 1).

When the versors are not units the tensors can be introduced as mere multipliers without affecting the versor conclusions.

29. We present one or two examples of quaternion division.

Ex. 1. *To express sin $(\theta + \phi)$ and cos $(\theta + \phi)$ in terms of sines and cosines of θ and ϕ.*

a, β, γ being unit vectors in the same plane (Fig. Art. 27), we have

$$\frac{\beta}{a} = \cos\theta + \epsilon \sin\theta,$$

$$\frac{\gamma}{\beta} = \cos\phi + \epsilon\sin\phi,$$

$$\frac{\gamma}{a} = \cos(\theta+\phi) + \epsilon\sin(\theta+\phi).$$

But $$\frac{\gamma}{a} = \frac{\gamma}{\beta} \cdot \frac{\beta}{a};$$

$\therefore \cos(\theta+\phi) + \epsilon\sin(\theta+\phi) = (\cos\theta + \epsilon\sin\theta)(\cos\phi + \epsilon\sin\phi)$;
whence multiplying out and equating, we have

$$\sin(\theta+\phi) = \sin\theta\cos\phi + \cos\theta\sin\phi,$$
$$\cos(\theta+\phi) = \cos\theta\cos\phi - \sin\theta\sin\phi.$$

Cor. If the action of the versors be in opposite directions, β lying beyond γ, we have (Art. 28)

$$\frac{a}{\gamma} = \cos(\theta-\phi) - \epsilon\sin(\theta-\phi).$$

But $$\frac{\beta}{\gamma} = \cos\phi + \epsilon\sin\phi,$$

$$\frac{a}{\beta} = \cos\theta - \epsilon\sin\theta;$$

$$\therefore \frac{a}{\gamma} = \frac{a}{\beta} \cdot \frac{\beta}{\gamma} \text{ gives}$$

$\cos(\theta-\phi) - \epsilon\sin(\theta-\phi) = (\cos\theta - \epsilon\sin\theta)(\cos\phi + \epsilon\sin\phi)$,
whence $\qquad \sin(\theta-\phi) = \sin\theta\cos\phi - \cos\theta\sin\phi,$
$$\cos(\theta-\phi) = \cos\theta\cos\phi + \sin\theta\sin\phi.$$

Ex. 2. *To find the cosine of the angle of a spherical triangle in terms of the sides.*

Let a, β, γ be unit vectors OA, OB, OC not in the same plane, then

$$\frac{\beta}{\gamma} = \frac{\beta}{a} \cdot \frac{a}{\gamma};$$

i.e. taking the scalar of each side,

$$\cos a = \cos c \cos b + S \cdot \left(V\frac{\beta}{a} \cdot V\frac{a}{\gamma} \right).$$

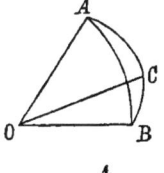

Now $SV\dfrac{\beta}{a}V\dfrac{a}{\gamma}$ is $\sin c \sin b \times$ cosine of the angle between perpendiculars to the planes AB, AC, and is therefore

$$\sin b \sin c \cos A \; ;$$

$$\therefore \; \cos a = \cos c \cos b + \sin c \sin b \cos A.$$

The reader will observe that in accordance with the results of Art. 21, the sign of the term involving $\cos A$ is $+$, seeing that it is in fact $-$ cosine (supplement of A).

Ex. 3. *The angles of a triangle are together equal to two right angles.*

What we shall prove in fact is that the exterior angles formed by producing the sides in the same direction are equal to four right angles.

Let unit vectors along BC, CA, AB be a, β, γ; and let the exterior angles formed by producing BC, CA, AB be θ, ϕ, ψ; then

$$\epsilon^{\frac{2\theta}{\pi}} a = \beta \;\; (27.\;1),$$

$$\epsilon^{\frac{2\phi}{\pi}} \beta = \gamma,$$

$$\epsilon^{\frac{2\psi}{\pi}} \gamma = a \; ;$$

$$\therefore \; \epsilon^{\frac{2\phi}{\pi}} . \epsilon^{\frac{2\theta}{\pi}} a = \epsilon^{\frac{2\phi}{\pi}} \beta = \gamma,$$

and

$$\epsilon^{\frac{2\psi}{\pi}} . \epsilon^{\frac{2\phi}{\pi}} . \epsilon^{\frac{2\theta}{\pi}} a = \epsilon^{\frac{2\psi}{\pi}} \gamma = a,$$

so that

$$\epsilon^{\frac{2\psi}{\pi}} . \epsilon^{\frac{2\phi}{\pi}} . \epsilon^{\frac{2\theta}{\pi}} = 1,$$

or

$$\epsilon^{\frac{2}{\pi}(\theta+\phi+\psi)} = 1 \;\; (27.\;2).$$

Hence (27. 3), $\dfrac{2}{\pi}(\theta+\phi+\psi)$ is an even multiple of 2. The first value is 4 ;

$$\therefore \; \theta + \phi + \psi = 2\pi,$$

or the exterior angles of a triangle are equal to four right angles.

It will be seen that the demonstration here given is of the nature of that given by Prof. Thomson in the Notes to his Euclid.

[More directly

$$\epsilon^{\frac{2A}{\pi}} \beta = -\gamma,$$

$$\epsilon^{\frac{2B}{\pi}} \gamma = -\alpha,$$

$$\epsilon^{\frac{2C}{\pi}} \alpha = -\beta.$$

From these

$$\epsilon^{\frac{2}{\pi}(A+B+C)} = -1,$$

or

$$A + B + C = \pi.]$$

Ex. 4. *In the figure of Euclid* I. 47 *the three lines* AL, BK, CF *meet in a point.*

Let $BC = \alpha$, $CA = \beta$, $AB = \gamma$; the sides being as usual denoted by a, b, c.

Let i be the vector which turns another negatively through a right angle in the plane of the paper, so that

$$BD = i\alpha, \quad CK = i\beta, \quad AG = i\gamma.$$

If BK, AL meet in O,

$$BO = xBK = x(\alpha + i\beta),$$

and

$$BO = BA + AO = BA + yBD$$

$$= -\gamma + yi\alpha;$$

$$\therefore \quad x(\alpha + i\beta) = -\gamma + yi\alpha,$$

$$xSa(\alpha + i\beta) = -Sa\gamma,$$

$$x = -\frac{Sa\gamma}{Sa(\alpha + i\beta)} = \frac{ac \cos B}{a^2 + ab \sin C}$$

$$= \frac{c^2}{a^2 + bc},$$

and

$$xSa\beta = ySia\beta;$$

$$\therefore \quad y = \frac{b}{c}x = \frac{bc}{a^2 + bc},$$

4—2.

which being symmetrical in b and c shews that CF, AL intersect in the same point in which BK, AL intersect.

COR.   Since $$\frac{BO}{BK} = \frac{c^2}{a^2 + bc},$$

we have $$\frac{CO}{CF} = \frac{b^2}{a^2 + bc};$$

also $$\frac{AO}{BD} = \frac{bc}{a^2 + bc};$$

$$\therefore \quad \frac{AO}{BD} + \frac{BO}{BK} + \frac{CO}{CF} = \frac{c^2 + b^2 + bc}{a^2 + bc} = 1.$$

Ex. 5.   *If ABCD be a quadrilateral inscribed in a circle;*

$$AB = \alpha, \ BC = \beta, \ CD = \gamma, \ DA = \delta;$$

then $$\alpha\beta\gamma = \frac{T\alpha T\beta T\gamma}{T\delta}\delta.$$

Let unit vectors along AB, BC, CD, DA be $\alpha', \beta', \gamma', \delta'$; and let the exterior angles at B and D be θ and ϕ respectively; then

$$\alpha'\beta'\gamma' = (-\cos\theta + \epsilon\sin\theta)\,\gamma' \ (21.\ 1)$$
$$= (\cos\phi + \epsilon\sin\phi)\,\gamma'$$
$$= \delta' \ (25.\ 1);$$

therefore, introducing the tensors,

$$\alpha\beta\gamma = \frac{T\alpha T\beta T\gamma}{T\delta}\,\delta.$$

Conjugate Quaternions.

30. If we designate by q the expression $-\cos\theta + \epsilon\sin\theta$, we have seen that it may be regarded as a *versor* through an angle θ in a certain direction. Now if we write $-\theta$ in place of θ in this expression it assumes the form $-\cos\theta - \epsilon\sin\theta$, which must on the same hypotheses be regarded a versor through the angle θ in the contrary direction.

When the quaternion is completed by the introduction of a tensor Tq, if we retain the same tensor to both forms of the versor, we have Sir W. Hamilton's *conjugate* quaternion defined thus :   The conjugate of a quaternion q, written Kq, has the same tensor, plane and angle as q has, only the angle is taken in the reverse way.

The analogy between q and Kq is precisely the same as that which exists between the two forms

$$R\left(\cos\phi + \sqrt{-1}\sin\phi\right) \text{ and } R\left(\cos\phi - \sqrt{-1}\sin\phi\right);$$

and as the product of the latter form is R^2, so the multiplication of the former produces $(Tq)^2$.

If we put $\qquad\qquad q = Sq + Vq,$

we shall have $\qquad\qquad Kq = Sq - Vq,$

and $\qquad\qquad qKq = (Sq)^2 + (TVq)^2,$

for $\qquad\qquad (Vq)^2 = -(TVq)^2,$ Art. 20.

It is almost self-evident that, since the change of order of multiplication of two vectors produces no other change than that of the sign of the vector part of the product (22),

$$K(qr) = KrKq,$$

q and r occurring in a changed order.

The following is a demonstration.

Let $\qquad\qquad q = Tq\left(-\cos\theta + a\sin\theta\right),$

$$r = Tr\left(-\cos\phi + \beta\sin\phi\right),$$

a and β being unit vectors; then

$$qr = TqTr\,(\cos\theta\cos\phi - a\sin\theta\cos\phi - \beta\cos\theta\sin\phi \\ + a\beta\sin\theta\sin\phi),$$

$$KrKq = TqTr\left(-\cos\phi - \beta\sin\phi\right)\left(-\cos\theta - a\sin\theta\right) \\ = TqTr\,(\cos\theta\cos\phi + a\sin\theta\cos\phi + \beta\cos\theta\sin\phi \\ + \beta a\sin\theta\sin\phi).$$

Now observing that βa has the same scalar part with $a\beta$, but the vector part with a contrary sign, we see that the two expressions for qr and for $KrKq$ likewise have the same scalar part, but that their vector parts have contrary signs.

Hence $\qquad\qquad K(qr) = KrKq.$

(See Tait, § 79 et sq.)

31. We propose, in this Article, to give and interpret one or two formulæ, relating to three or more vectors, which are indispensable to our progress, reserving to a separate Chapter the demonstration and application of other formulæ, the value of which the reader can hardly as yet be expected to understand.

1. To express $S.\,\alpha\beta\gamma$ geometrically.

First suppose α, β, γ to be unit vectors OA, OB, OC.

Let $AOB = \theta$, and the angle which OC makes with the plane $AOB = \phi$; then since

$$\alpha\beta = -\cos\theta + \epsilon\sin\theta \text{ (Art. 21)},$$

where ϵ is perpendicular to the plane AOB,

$$S.\,\alpha\beta\gamma = S(-\cos\theta + \epsilon\sin\theta)\gamma$$
$$= S\epsilon\gamma\sin\theta.$$

Now $S\epsilon\gamma = -\cos.$ angle between ϵ and γ

$\qquad\qquad = -\sin.$ angle between plane AOB
$\qquad\qquad\qquad\qquad$ and OC

$\qquad\qquad = -\sin\phi\,;$

$\therefore\ S.\,\alpha\beta\gamma = -\sin\phi\sin\theta.$

Next if α, β, γ are not units, but have respectively the lengths $T\alpha$, $T\beta$, $T\gamma$, or a, b, c; we shall have

$$S.\,\alpha\beta\gamma = -abc\sin\theta\sin\phi.$$

But $ab\sin\theta$ is the area of the parallelogram of which the adjacent sides are a, b; and $c\sin\phi$ is the perpendicular from C on the plane of the parallelogram;

$\therefore\ -S.\,\alpha\beta\gamma = ab\sin\theta\,.\,c\sin\phi$

$\qquad\qquad\qquad = $ volume of parallelepiped of which three conterminous edges are OA, OB, OC.

2.   From the nature of the case, no change of order amongst the vectors α, β, γ can make any change in the value (apart from the sign) of the scalar of the product of the three vectors; for it will in every case produce the volume of the same parallelepiped.

$$\therefore\ S . \alpha\beta\gamma = \pm S . \gamma\alpha\beta = \pm S . \alpha\gamma\beta,\ \&c.$$

COR. 1.   The volume of the triangular pyramid, of which OA, OB, OC are conterminous edges is $-\dfrac{1}{6} S . \alpha\beta\gamma$.

COR. 2.   If α, β, γ are in the same plane, $\phi = 0$;

$$\therefore\ S . \alpha\beta\gamma = 0.$$

Conversely, if $S . \alpha\beta\gamma = 0$, none of the vectors α, β, γ being themselves 0, we must have either $\theta = 0$ or $\phi = 0$; hence in either case the three vectors are co-planar.

3.   Since $V\alpha\beta = \gamma'$ (21. 3), a vector perpendicular to the plane OAB (fig. of formula 2); $V\beta\gamma = \alpha'$, a vector perpendicular to the plane OBC; and since γ', α' are both perpendicular to OB, the line along which is the vector β; OB is perpendicular to the plane which passes through γ', α', and therefore (21. 3) is in the direction of $V\gamma'\alpha'$; hence

$$V(V\alpha\beta\, V\beta\gamma) = V\gamma'\alpha' - m\beta,$$

or the vector of the product of two resultant vectors, one of the constituents of each of which is the same vector, is a multiple of that vector.

4.   If $OA = \alpha$, $OB = \beta$, $OD = \delta$, $OE = \epsilon$; and if the planes OAB, ODE intersect in OP; it follows, as in (3), that, $V\alpha\beta$ and $V\delta\epsilon$ being both perpendicular to OP,

$$V(V\alpha\beta\, V\delta\epsilon) \text{ is along } OP \text{ and is therefore } = nOP.$$

5.   *Connection between the representation of the position of a point by a vector and its representation by Cartesian co-ordinates.*

If x, y, z be the perpendicular distances of a point P in space from the planes of yz, zx, xy respectively (fig. of Art. 16); i, j, k

unit vectors in the directions of x, y, z; then xi is the *vector* of which the line is x (Art. 3); consequently OM along Ox, MN parallel to Oy and NP parallel to Oz, being x, y, z as co-ordinates, they are xi, yj, zk as vectors.

Now   vector $OP = OM + MN + NP$,

and is therefore $\rho = xi + yj + zk$.

The same method of representation is evidently applicable when the planes of reference are not mutually at right angles. If x, y, z be the co-ordinates of P referred to oblique co-ordinates; α, β, γ unit vectors parallel respectively to x, y, z; then

$$\text{vector } OP = x\alpha + y\beta + z\gamma.$$

Cor.   When x, y, z are at right angles to one another

$$\rho = xi + yj + zk$$

gives $$Si\rho = -x, \quad Sj\rho = -y, \quad Sk\rho = -z;$$

$$\therefore (Si\rho)^2 + (Sj\rho)^2 + (Sk\rho)^2 = x^2 + y^2 + z^2$$
$$= OP^2.$$

Ex.   *To find the volume of the pyramid of which the vertex is a given point and the base the triangle formed by joining three given points in the rectangular co-ordinate axes.*

Let A, B, C be the three given points;

$$\text{line } OA = a, \quad OB = b, \quad OC = c;$$

x, y, z the co-ordinates of the given point P,

then   vector $OA = ai$, $OB = bj$, $OC = ck$;

and $OP = xi + yj + zk$;

$$\therefore PA = OA - OP = -\{(x-a)\,i + yj + zk\},$$
$$PB = -\{xi + (y-b)\,j + zk\},$$
$$PC = -\{xi + yj + (z-c)\,k\}.$$

Now the volume of the pyramid $PABC$ is

$$-\frac{1}{6} S (PA \cdot PB \cdot PC) \quad (31.\,2.\text{ Cor. 1})$$

$$= -\frac{1}{6} S \cdot \{(x-a)\,i + yj + zk\}\,\{xi + (y-b)\,j + zk\}\,\{xi + yb + (z-c)\,k\}.$$

Multiplying out and observing that only terms which involve all of the three vectors i, j, k produce a scalar in the product, we get

$$(+\text{ or }-)\text{ Vol.} = -\frac{1}{6}\{(x-a)\,(bz+cy-bc)-cxy-bxz\}$$

$$= \frac{1}{6}\,abc\left(\frac{x}{a}+\frac{y}{b}+\frac{z}{c}-1\right).$$

The *sign* of the result will of course depend on the position of P.

ADDITIONAL EXAMPLES TO CHAP. III.

1.   If in the figure of Euclid I. 47 DF, GH, KE be joined, the sum of the squares of the joining lines is three times the sum of the squares of the sides of the triangle.

The same is true whatever be the angle A.

2.   Prove that

$$4AD^2 \text{ (Art. 7, Ex. 4)} = 2\,(AB^2 + AC^2) - BC^2.$$

3.   If P, Q, R, S be points in the sides AB, BC, CD, DA of a rectangle, such that $PQ = RS$, prove that

$$AR^2 + CS^2 = AQ^2 + CP^2.$$

4.   The sum of the squares of the three sides of a triangle is equal to three times the sum of the squares of the lines drawn from the angles to the mean point of the triangle.

5.   In any quadrilateral, the product of the two diagonals and the cosine of their contained angle is equal to the sum or difference of the two corresponding products for the pairs of opposite sides.

6.   If a, b, c be three conterminous edges of a rectangular parallelepiped; prove that four times the square of the area of the triangle which joins their extremities is

$$= a^2 b^2 + b^2 c^2 + c^2 a^2.$$

7.   If two pairs of opposite edges of a tetrahedron be respectively at right angles, the third pair will be also at right angles.

8.   Given that each edge of a tetrahedron is equal to the edge opposite to it.   Prove that the lines which join the points of bisection of opposite edges are at right angles to those edges.

9.   If from the vertex O of a tetrahedron $OABC$ the straight line OD be drawn to the base making equal angles with the faces OAB, OAC, OBC; prove that the triangles OAB, OAC, OBC are to one another as the triangles DAB, DAC, DBC.

CHAPTER IV.

THE STRAIGHT LINE AND PLANE.

32. EQUATIONS of a straight line.

1. Let β be a vector (unit or otherwise) parallel to or along the straight line; α the vector to a given point A in the line, ρ that to any point whatever P in the line, starting from the same origin O; then AP is a vector parallel to β

$$= x\beta, \text{ say,}$$

and $\qquad\qquad OP = OA + AP$

gives $\qquad\qquad \rho = \alpha + x\beta \quad (1)$

as the equation of the line.

2. Another form in which the equation of a straight line may be expressed is this: let $OA = \alpha$, $OB = \beta$ be the vectors to two given points in the line; then

$$AB = \beta - \alpha \text{ and } AP = x\,(\beta - \alpha)\,;$$
$$\therefore \ \rho = \alpha + x\,(\beta - \alpha) \ (2).$$

Of course the β of No. 2 is not that of No. 1. The first form of the equation supposes the direction of the line and the position of one point in it to be given, the second form supposes two points in it to be given.

3. A third form may be exhibited in which the perpendicular on the line from the origin is given.

Let OD perpendicular to $AP = \delta$; then

$$DP = \rho - \delta \text{ and } S\delta(\rho - \delta) = 0,$$

because OD is perpendicular to AP (22. 7);

$$\text{i.e. } S\delta\rho = C \quad (3),$$

where C is a constant.

(*Note*. In addition to this we must have the equation of the plane of the paper, in which ρ is tacitly supposed to lie. This may be written as $S\epsilon\rho = 0$.)

33. Equation of a plane.

Let P be any point in the plane, OD perpendicular to the plane; and let

$$OD = \delta, \quad OP = \rho;$$

then $$\rho - \delta = DP,$$

which is in a direction perpendicular to OD;

$$\therefore S\delta(\rho - \delta) = 0,$$

$$\text{or } S\delta\rho = \delta^2,$$

$$\text{or } S\frac{\rho}{\delta} = 1.$$

COR. 1. If $S\delta\rho = C$ be the equation of a plane, δ is a vector in the direction perpendicular to the plane.

COR. 2. If the plane pass through O, ρ can have the value zero,

$$\therefore S\delta\rho = 0 \text{ is the equation.}$$

COR. 3. Since a vector can be drawn in the plane through D, parallel to any given vector in or parallel to the plane; if β be any vector in or parallel to the plane, $S\delta\beta = 0$.

34. We proceed to exhibit certain modifications of the equations of a straight line and plane, and one or two results immediately deducible from the forms of those equations.

1. To find the equation of a straight line which is perpendicular to each of two given straight lines.

Let β, γ be vectors from a given point A in the required line, and parallel respectively to the given lines.

If $OA = a$ as before, then since (22. 8) $V\beta\gamma$ is a vector along the line whose equation is required; we have

$$\rho - a = xV\beta\gamma,$$
$$\text{or } \rho = a + xV\beta\gamma,$$

as the equation of the line.

2.   To find the length of the perpendicular from the origin on a given line.

Equation (1) of Art. 32 is

$$\rho = a + x\beta.$$

If now
$$\rho = OD = \delta ;$$

we get
$$S\delta^2 = S\delta a,$$

or
$$- OD^2 = S\delta a ;$$

$$\therefore OD = -\frac{S\delta a}{OD} = - Sa\,U\delta,$$

$U\delta$ being the unit vector perpendicular to the line.

Cor.   The same result is true of a plane.

3.   To find the length of the perpendicular from a given point on a given plane.

Let $Sa\rho = C$ be the equation of the plane, γ the vector to the given point.

Then if the vector perpendicular be xa (33. Cor. 1),

$$\rho = \gamma + xa$$

gives
$$Sa\gamma + xa^2 = C,$$

and the vector perpendicular is

$$xa = + a^{-1}(C - Sa\gamma) ;$$

the square of which with a − sign is the square of the perpendicular.

4.   To find the length of the common perpendicular to each of two given straight lines.

Let β, β_1 be unit vectors along the lines; α, α_1 vectors to given points in the lines;

$$\rho = \alpha + x\beta,$$
$$\rho_1 = \alpha_1 + x_1\beta_1,$$

the vectors to the extremities of the common perpendicular δ.

Then since δ is perpendicular to both lines, it is perpendicular to the plane which passes through two straight lines drawn parallel to them through a given point;

$$\therefore (21.\ 3)\ \delta = y V\beta\beta_1.$$

But $\qquad \delta = \rho - \rho_1 = \alpha + x\beta - \alpha_1 - x_1\beta_1,$

hence $\qquad S . \delta\beta\beta_1 = S . (\alpha - \alpha_1) \beta\beta_1 ;$

i.e. $S(y V\beta\beta_1 . \beta\beta_1) = S . (\alpha - \alpha_1) \beta\beta_1,$

or $\quad y (V\beta\beta_1)^2 = S . (\alpha - \alpha_1) \beta\beta_1,$

because $\qquad\qquad S V\beta\beta_1 S\beta\beta_1 = 0 ;$

$$\therefore\ y = \frac{S . (\alpha - \alpha_1) \beta\beta_1}{(V\beta\beta_1)^2},$$

whence $\delta = y V\beta\beta_1$ is known.

5. To find the equation of a plane which passes through three given points.

Let α, β, γ be the vectors of the points.

Then $\rho - \alpha$, $\alpha - \beta$, $\beta - \gamma$ are in the same plane.

$\therefore$ (Art. 31. 2. Cor. 2) $S . (\rho - \alpha)(\alpha - \beta)(\beta - \gamma) = 0,$

or $\qquad S\rho (V\alpha\beta + V\beta\gamma + V\gamma\alpha) - S . \alpha\beta\gamma = 0$

is the equation required.

Cor. $V\alpha\beta + V\beta\gamma + V\gamma\alpha$ is a vector in the direction perpendicular to the plane; therefore (No. 3) the perpendicular vector from the origin

$$= S . \alpha\beta\gamma . (V\alpha\beta + V\beta\gamma + V\gamma\alpha)^{-1}.$$

6. To find the equation of a plane which shall pass through a given point and be parallel to each of two given straight lines.

Let γ be the vector to the given point, $\rho = a + x\beta$, $\rho = a_1 + x_1\beta_1$ the lines; then if lines be drawn in the required plane parallel to each of the given straight lines—these lines as vectors will be β, β_1: also $\rho - \gamma$ is a vector line in the plane;

$$\therefore \; S \cdot \beta\beta_1 (\rho - \gamma) = 0 \; (31. \; 2. \; \text{Cor. 2}),$$

which is the equation required.

7. To find the equation of a plane which shall pass through two given points and be perpendicular to a given plane.

Let a, β be the vectors to the given points, $S\delta\rho = C$ the equation of the plane; then the three lines $\rho - a$, $a - \beta$, δ are vectors in the plane;

$$\therefore \; S \cdot (\rho - a)(a - \beta)\delta = 0,$$

$$\text{or} \; S \cdot \rho (a - \beta)\delta + S \cdot a\beta\delta = 0.$$

8. *To find the condition that four points shall be in the same plane.*

1. Let OA, OB, OC, OD or a, β, γ, δ be the vectors to the four points; then $\delta - a$, $\delta - \beta$, $\delta - \gamma$ are vectors in the same plane;

$$\therefore \; S \cdot (\delta - a)(\delta - \beta)(\delta - \gamma) = 0 \; (31. \; 2. \; \text{Cor. 2}),$$

$$\text{or} \; S \cdot \delta\beta\gamma + S \cdot a\delta\gamma + S \cdot a\beta\delta = S \cdot a\beta\gamma \; (1).$$

2. Another form of the condition is to be obtained by assuming that

$$d\delta + c\gamma + b\beta + aa = 0 \; (2),$$

and substituting in equation (1) the value of δ deduced from this equation. The result is

$$\frac{a}{d} + \frac{b}{d} + \frac{c}{d} + 1 = 0,$$

$$\text{or} \; a + b + c + d = 0 \; (3).$$

Equation (1), or the concurrence of equations (2) and (3) is the condition necessary and sufficient for coplanarity.

9. To find the line of intersection of two planes through the origin.

Let $Sa\rho = 0$, $S\beta\rho = 0$ be the planes.

Since every line in the one plane is perpendicular to a; and every line in the other perpendicular to β; the line required is perpendicular to both a and β, and is therefore parallel to $Va\beta$, or $\rho = xVa\beta$ is the equation.

10. The equation of the plane which passes through O and the line of intersection of the planes $Sa\rho = a$, $S\beta\rho = b$ is

$$Sp\,(a\beta - ba) = 0.$$

For 1° it is a plane through O; 2° if ρ be such that $Sa\rho = a$, then must $S\beta\rho = b$.

11. To find the equation of the line of intersection of the two planes.

Let $$\rho = ma + n\beta + xVa\beta$$

be the equation required.

Then $$Sa\rho = ma^2 + nSa\beta = a,$$

since $Va\beta$ is perpendicular to a, and similarly

$$S\beta\rho = mSa\beta + n\beta^2 = b,$$

$$\therefore \; m = \frac{a\beta^2 - bSa\beta}{a^2\beta^2 - (Sa\beta)^2} = \frac{bSa\beta - a\beta^2}{(Va\beta)^2} \;\; (\text{Art. 22. 9}),$$

$$n = \frac{aSa\beta - ba^2}{(Sa\beta)^2 - a^2\beta^2} = \frac{aSa\beta - ba^2}{(Va\beta)^2}.$$

35. We offer a few simple examples.

Ex. 1. *To find the locus of the middle points of all straight lines which are terminated by two given straight lines.*

Let AP, BQ be the two given straight lines, unit vectors parallel to which are β, γ; AB the line which is perpendicular to both AP, BQ.

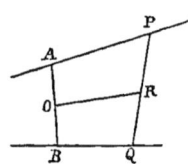

Let O be the middle point of AB; vector $OA = a$; R the middle point of any line PQ, vector $OR = \rho$; then

$$OP = \rho + RP = a + x\beta,$$
$$OQ = \rho + RQ = -a + y\gamma.$$

But
$$RP + RQ = 0 ;$$
$$\therefore\ 2\rho = x\beta + y\gamma ;$$

hence, since
$$Sa\beta = 0,\ Sa\gamma = 0,$$

$Sa\rho = 0$ is the equation required; and the locus is a plane passing through O (33. Cor. 2), and perpendicular to OA (33. Cor. 1).

Note that, if $\beta \parallel \gamma$, we have simply

$$2\rho = x'\beta ;$$

and, as there is now but one scalar indeterminate, the locus is a straight *line* instead of a *plane*.

Ex. 2. *Planes cut off, from the three rectangular co-ordinate axes, pyramids of equal volume, to find the locus of the feet of perpendiculars on them from the origin.*

Here the axes are *given*, so that i, j, k are known unit vectors.

Let ai, bj, ck be the portions cut off from the axes by a plane, the perpendicular on which from the origin is ρ.

Then $\rho - ai$ is perpendicular to ρ ;

$$\therefore\ S\rho\,(\rho - ai) = 0,$$
$$\text{or }\ \rho^2 = aSi\rho.$$

Similarly,
$$\rho^2 = bSj\rho,$$
$$\rho^2 = cSk\rho.$$

Hence
$$\rho^6 = abc\ Si\rho\ Sj\rho\ Sk\rho$$
$$= C\ Si\rho\ Sj\rho\ Sk\rho,$$

since abc is by the problem constant.

If x, y, z be the co-ordinates of ρ this equation gives at once

$$(x^2 + y^2 + z^2)^3 = Cxyz$$

as the equation required.

5

Ex. 3. *To find the locus of the middle points of straight lines terminated by two given straight lines and all parallel to a given plane.*

Retaining the figure and notation of Ex. 1, let δ be the vector perpendicular to the given plane : we have

$$2\rho = x\beta + y\gamma,$$

$$2QP = 2a + x\beta - y\gamma.$$

Now
$$S\delta QP = 0 \quad (33. \text{ Cor. } 3) ;$$

$$\therefore \ S\delta \,(2a + x\beta - y\gamma) = 0 ;$$

$$\therefore \ y = \frac{2Sa\delta}{S\gamma\delta} + x \,\frac{S\beta\delta}{S\gamma\delta},$$

and
$$2\rho = x\beta + \frac{2Sa\delta}{S\gamma\delta}\,\gamma + x\,\frac{S\beta\delta}{S\gamma\delta}\,\gamma$$

$$= a\gamma + x\,(\beta + b\gamma),$$

where $a = \dfrac{2Sa\delta}{S\gamma\delta}$, $b = \dfrac{S\beta\delta}{S\gamma\delta}$ are constants ; ($S\gamma\delta$ for instance is the negative of the cosine of the angle between one of the given lines and the perpendicular to the given plane).

Now $\beta + b\gamma$ is a known vector lying between β and γ; call it ϵ, and $2\rho = a\gamma + x\epsilon$ is the equation required; which is that of a straight line, not generally passing through O (32. 1).

Ex. 4. *OA, OB are two fixed lines, which are cut by lines AB, A′B′ so that the area AOB is constant; and also the product OA, OA′ constant. It is required to find the locus of the intersections of AB, A′B′.*

Let the unit vectors along OA, OB be a, β respectively.

$$OA = ma, \quad OA' = m'a,$$

$$OB = n\beta, \quad OB' = n'\beta ;$$

then the conditions of the problem are

$$mn = m'n' = C,$$

$$mm' = a.$$

Now if AB, $A'B'$ intersect in P, and $OP = \rho$, we have

$$\rho = OA + AP$$

$$= ma + x\,(n\beta - ma),$$

$$\rho = OA' + A'P$$

$$= m'a + x'\,(n'\beta - m'a)\,;$$

or $\rho = ma + x\left(\dfrac{C}{m}\beta - ma\right),$

$$\rho = m'a + x'\left(\dfrac{C}{m'}\beta - m'a\right)\,;$$

$$\therefore\ \ m - xm = m' - x'm',$$

and $\qquad\qquad \dfrac{x}{m} = \dfrac{x'}{m'}.$

Hence $\qquad\qquad x = \dfrac{m}{m + m'}$

$$= \dfrac{m^2}{m^2 + a},$$

$$1 - x = \dfrac{a}{m^2 + a},$$

and $\quad \rho = \dfrac{m}{m^2 + a}\,(aa + C\beta),$

and the locus required is a straight line, the diagonal of the parallelogram whose sides are aa, $C\beta$.

Ex. 5.   *To find the locus of a point such that the ratio of its distances from a given point and a given straight line is constant—all in one plane.*

Let S be the given point, DQ the given straight line, $SP = ePQ$ the given relation.

Let   vector $SD = a$, $SP = \rho$, $DQ = y\gamma$, γ being the unit vector along DQ,

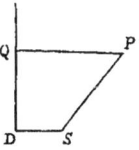

$$PQ = xa\,;$$

then $\qquad\qquad T\rho = eT\,(PQ),$

gives $\qquad$ $\rho^2 = e^2 PQ^2$, where PQ is a vector,

$$= e^2 (xa)^2$$

$$= e^2 x^2 a^2.$$

But $\qquad$ $\rho + xa = SQ = SD + DQ$

$$= a + y\gamma \,;$$

$$\therefore Sa\rho + xa^2 = a^2, \text{ for } Sa\gamma = 0 \,;$$

and $\qquad$ $x^2 a^4 = (a^2 - Sa\rho)^2 \,;$

hence $\qquad$ $a^2 \rho^2 = e^2 (a^2 - Sa\rho)^2,$

a surface of the second order, whose intersection with the plane $S \cdot a\gamma\rho = 0$ is the required locus.

Ex. 6. *The same problem when the points and line are not in the same plane.*

Retaining the same figure and notation, we see that PQ is no longer a multiple of a; but

$$PQ = SQ - SP$$

$$= a + y\gamma - \rho \,;$$

$$\therefore \rho^2 = e^2 (a + y\gamma - \rho)^2,$$

and because PQ is perpendicular to DQ

$$S\gamma (a + y\gamma - \rho) = 0 \,;$$

$$\therefore (y\gamma^2, \text{ i.e.}) - y = S\gamma\rho,$$

and $\qquad$ $\rho^2 = e^2 (a - \gamma S\gamma\rho - \rho)^2,$

a surface of the second order.

Cor. If $e = 1$, and the surface be cut by a plane perpendicular to DQ whose equation is $S\gamma\rho = c$, the equation of the section is

$$a^2 + c^2 - 2Sa\rho = 0,$$

another plane, so that the section is a straight line.

Ex. 7. *To find the locus of the middle points of lines of given length terminated by each of two given straight lines.*

Retaining the figure and notation of Ex. 1, and calling $RP\ c$, we have

$$2\rho = x\beta + y\gamma \ (1),$$

and $$2RP = RP - RQ = 2a + x\beta - y\gamma \ (2).$$

From equation (1) we have

$$Sa\rho = 0 \ (22.\ 7),$$
$$2S\beta\rho = -x + yS\beta\gamma,$$

because β is a unit vector,

$$2S\gamma\rho = xS\beta\gamma - y.$$

The first of these three equations shews that ρ lies in a plane through O perpendicular to AB (33. Cor. 2).

The second and third equations give

$$x = \frac{2\,(S\beta\rho + S\beta\gamma\,S\gamma\rho)}{(S\beta\gamma)^2 - 1},$$
$$y = \frac{2\,(S\gamma\rho + S\beta\gamma\,S\beta\rho)}{(S\beta\gamma)^2 - 1}.$$

Now (2) gives, by squaring,

$$-4c^2 = 4a^2 + x^2\beta^2 + y^2\gamma^2 - 2xy\,S\beta\gamma,$$

in which, if the values of x and y just obtained be substituted, there results an equation of the second order in ρ.

Hence the locus required is a plane curve of the second order, or a conic section, which by the very nature of the problem must be finite in extent and therefore an ellipse.

Ex. 8.   *If a plane be drawn through the points of bisection of two opposite edges of a tetrahedron it will bisect the tetrahedron.*

Let D, E be the middle points of OB, AC : $DFEG$ the cutting plane : OA, OB, $OC = a$, β, γ respectively.

$$OG = m\gamma, \quad AF = n\,(\beta - a).$$

The portion $ODGEA$ consists of three tetrahedra whose common vertex is O, and bases the triangles AEF, EFG, FGD.

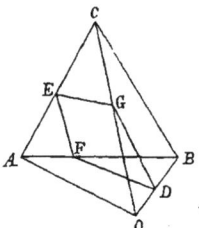

Now $$OE = \frac{1}{2}\,(\gamma + a),$$

$$OD = \frac{1}{2}\,\beta,$$

$$OG = m\gamma,$$

$$OF = a + n\,(\beta - a)\,;$$

and 6 times the volume cut off

$$= S\,.\,a\,\frac{1}{2}\,(a + \gamma)\{a + n\,(\beta - a)\}$$

$$+ S\,.\,\frac{1}{2}\,(a + \gamma)\,m\gamma\,\{a + n\,(\beta - a)\}$$

$$+ S.\{a + n\,(\beta - a)\}\,m\gamma\,\frac{1}{2}\,\beta \quad (31.\ 2\ \text{Cor.}\ 1)$$

$$= \frac{1}{2}\,\{n + nm + (1 - n)\,m\}\,S\,.\,a\gamma\beta$$

$$= \frac{1}{2}\,(n + m)\,S\,.\,a\gamma\beta.$$

But since E, G, D, F are in one plane, and

$$2m\,(1 - n)\,OE - (1 - n)\,OG + 2mn\,OD - m\,OF = 0,$$

we must have (34. 8)

$$2m\,(1 - n) - (1 - n) + 2mn - m = 0\,;$$

$$\therefore\ m + n = 1\,;$$

and 6 times the whole volume cut off

$$= \frac{1}{2}\,S\,.\,a\gamma\beta$$

$$= \frac{1}{2}\ \text{of 6 times the whole volume,}$$

hence the plane bisects the tetrahedron.

Cor. The plane cuts other two edges at F and G, so that

$$\frac{AF}{AB} + \frac{OG}{OC} = 1.$$

ADDITIONAL EXAMPLES TO CHAP. IV.

1. Straight lines are drawn terminated by two given straight lines, to find the locus of a point in them whose distances from the extremities have a given ratio.

2. Two lines and a point S are given, not in one plane ; find the locus of a point P such that a perpendicular from it on one of the given lines intersects the other, and the portion of the perpendicular between the point of section and P bears to SP a constant ratio. Prove that the locus of P is a surface of the second order.

3. Prove that the section of this surface by a plane perpendicular to the line to which the generating lines are drawn perpendicular is a circle.

4. Prove that the locus of a point whose distances from two given straight lines have a constant ratio is a surface of the second order.

5. A straight line moves parallel to a fixed plane and is terminated by two given straight lines not in one plane ; find the locus of the point which divides the line into parts which have a constant ratio.

6. Required the locus of a point P such that the sum of the projections of OP on OA and OB is constant.

7. If the sum of the perpendiculars on two given planes from the point A is the same as the sum of the perpendiculars from B, this sum is the same for every point in the line AB.

8. If the sum of the perpendiculars on two given planes from each of three points A, B, C (not in the same straight line) be the same, this sum will remain the same for every point in the plane ABC.

9. A solid angle is contained by four plane angles. Through a given point in one of the edges to draw a plane so that the section shall be a parallelogram.

10. Through each of the edges of a tetrahedron a plane is drawn perpendicular to the opposite face. Prove that these planes pass through the same straight line.

11. ABC is a triangle formed by joining points in the rectangular co-ordinates OA, OB, OC; OD is perpendicular to ABC. Prove that the triangle AOB is a mean proportional between the triangles ABC, ABD.

12. $V\sigma\rho V\beta\rho + (V\alpha\beta)^2 = 0$ is the equation of a hyperbola in ρ, the asymptotes being parallel to α, β.

CHAPTER V.

THE CIRCLE AND SPHERE.

36. *Equations of the circle.*

Let AD be the diameter of the circle, centre C, radius $= a$, P any point.

If vector $CD = a$, $CP = \rho$,

we have $\rho^2 = -a^2 \ldots\ldots\ldots(1)$.

If however $AP = \rho$,

$$CP = \rho - a,$$

we have $(\rho - a)^2 = -a^2 \ldots\ldots\ldots (2)$.

If O be any point,

$$OP = \rho, \quad OC = \gamma, \quad CP = \rho - \gamma,$$

we have $(\rho - \gamma)^2 = -a^2 \ldots\ldots\ldots\ldots\ldots(3)$.

These are the three forms of the vector equation.

Form (2) may be written

$$\rho^2 - 2Sa\rho = 0.$$

If $OC = c$, form (3) may be written

$$\rho^2 - 2S\gamma\rho = c^2 - a^2.$$

EXAMPLES.

37. Ex. 1. *The angle in a semicircle is a right angle.*

Taking the second form

$$\rho^2 - 2Sa\rho = 0,$$

we may again write it

$$S\rho \, (\rho - 2a) = 0 \, ;$$

therefore ρ, $\rho - 2a$ are vectors at right angles to one another. But $\rho - 2a$ is DP ;

$$\therefore DPA \text{ is a right angle.}$$

Ex. 2. *If through any point O within or without a circle, a straight line be drawn cutting the circle in the points P, Q, the product $OP \cdot OQ$ is always the same for that point.*

The third form of the equation may be written

$$(T\rho)^2 + 2T\rho S\gamma U\rho + c^2 - a^2 = 0,$$

which shews that $T\rho$ has two values corresponding to each value of $U\rho$, the product of which is $c^2 - a^2$. Therefore, &c.

Ex. 3. *If two circles cut one another, the straight line which joins the points of section is perpendicular to the straight line which joins the centres.*

Let O, C be the centres, P, Q the points of section ;

vector $OC = a$; a, b the radii ;

then (as vectors)

$$OP^2 = -a^2,$$

$$(OP - a)^2 = -b^2 ;$$

$$\therefore SaOP = C, \text{ a constant.}$$

Similarly, $SaOQ = C$, the same constant ;

$$\therefore Sa\,(OQ - OP) = 0,$$

$$\text{or } SaPQ = 0,$$

i.e. PQ is at right angles to OC.

Ex. 4. *O is a fixed point, AB a given straight line. A point Q is taken in the line OP drawn to a point P in AB, such that*

$$OP \cdot OQ = k^2 ;$$

to find the locus of Q.

Let OA perpendicular to AB be a, vector a ;

$$OQ = \rho, \quad OP = x\rho ;$$

then $T(OP \cdot OQ) = k^2,$

or $x\rho^2 = -k^2.$

But
$$Sa\left(x\rho - a\right) = 0\ ;$$
$$\therefore\ xSa\rho = -a^2\ ;$$

hence
$$\rho^2 = \frac{k^2}{a^2}Sa\rho$$

is the equation of the locus of Q, which is therefore a circle, passing through O.

Ex. 5. *Straight lines are drawn through a fixed point, to find the locus of the feet of perpendiculars on them from another fixed point.*

Let O, A be the points, the lines being drawn through A. Let $OA = a$, and let $\rho = a + x\beta$ be the equation of one of the lines through A, δ the perpendicular on it from O.

Then
$$\delta = a + x\beta,$$
and
$$S\delta^2 = Sa\delta,$$
because δ is perpendicular to β ;

i.e. $\delta^2 - Sa\delta = 0$,

the equation of a circle whose diameter is OA.

Ex. 6. *A chord QR is drawn parallel to the diameter AB of a circle: P is any point in AB; to prove that*
$$PQ^2 + PR^2 = PA^2 + PB^2.$$

Let
$$CQ = \rho,\ \ CR = \rho',\ \ PC = a\ ;$$
then
$$PQ^2 = -\left(\text{vector } PQ\right)^2$$
$$= -\left(a + \rho\right)^2 = -\left(a^2 + 2Sa\rho + \rho^2\right),$$
$$PR^2 = -\left(a + \rho'\right)^2 = -\left(a^2 + 2Sa\rho' + \rho'^2\right)\ ;$$
$$\therefore\ PQ^2 + PR^2 = 2PC^2 + 2AC^2 - 2\left(Sa\rho + Sa\rho'\right).$$

But
$$S\left(\rho + \rho'\right)\left(\rho - \rho'\right) = 0 \text{ and } \rho - \rho' = xa,$$
because QR is parallel to AB ;

$$\therefore\ Sa\rho + Sa\rho' = 0,$$
and
$$PQ^2 + PR^2 = 2PC^2 + 2AC^2$$
$$= PA^2 + PB^2.$$

Ex. 7. *If three given circles be cut by any other circle, the chords of section will form a triangle, the loci of the angular points of which are three straight lines respectively perpendicular to the lines which join the centres of the given circles ; and these three lines meet in a point.*

Let A, B, C be the centres of the three given circles; a, b, c their radii; a, β, γ the vectors to A, B, C from the origin O; OA, OB, OC respectively p, q, r; D the centre of the cutting circle whose radius is R, $OD = s$, vector $OD = \delta$, ρ the vector to a point of section of circle D with circle A ; then we shall have

$$(\rho - a)^2 = -a^2, \quad (\rho - \delta)^2 = -R^2,$$
$$\text{and} \therefore \ 2S(\delta - a)\rho = R^2 - a^2 - s^2 + p^2.$$

Now this is satisfied by the values of ρ to both points of section ; and being the equation of a straight line (32. 3) is the equation of the line joining the points of section of circle D with circle A—call it line 1, and so of the others ; then

$$\text{line 1 is } 2S(\delta - a)\rho = R^2 - a^2 - s^2 + p^2,$$
$$\text{line 2 is } 2S(\delta - \beta)\rho' = R^2 - b^2 - s^2 + q^2,$$
$$\text{line 3 is } 2S(\delta - \gamma)\rho'' = R^2 - c^2 - s^2 + r^2.$$

If 1 and 2 intersect in P whose vector is ρ_1, 1 and 3 in Q (ρ_2); 2 and 3 in R (ρ_3), we shall have by subtraction

$$\text{at } P, \ 2S(a - \beta)\rho_1 = a^2 - b^2 - p^2 + q^2 ;$$
$$\text{at } Q, \ 2S(\gamma - a)\rho_2 = -a^2 + c^2 + p^2 - r^2 ;$$
$$\text{at } R, \ 2S(\beta - \gamma)\rho_3 = b^2 - c^2 - q^2 + r^2 ;$$

therefore (32. 3) the loci of P, Q, R are straight lines, perpendicular respectively to AB, AC, BC.

Also at the point of intersection of the first and third of these lines, we have, by addition,

$$2S(a - \gamma)\rho = a^2 - c^2 - p^2 + r^2,$$

which is satisfied by the second : hence the three loci meet in a point.

Ex. 8. *To find the equation of the cissoid.*

AQ is a chord in a circle whose diameter is AB, QN perpendicular to AB.

AM is taken equal to BN, and MP is drawn perpendicular to AB to meet AQ in P; the locus of P is the cissoid.

Let vector $AP = \pi$, $AC = a$, $AM = ya$, $AQ = x\pi$;

then $\qquad y : 1 :: 2 - y : x$, by the construction;

$$\therefore\ y = \frac{2}{1+x}.$$

Now $\qquad x^2\pi^2 - 2xSa\pi = 0$

is the equation of the circle;

$$\therefore\ x = \frac{2Sa\pi}{\pi^2}.$$

Also $\qquad \pi = AM + MP$

$$= ya + \gamma\,;$$

$$\therefore\ Sa\pi = ya^2,$$

$$y = \frac{Sa\pi}{a^2}\,;$$

hence $\qquad \left(1 + \frac{2Sa\pi}{\pi^2}\right)\frac{Sa\pi}{a^2} = 2,$

and $\qquad (\pi^2 + 2Sa\pi)\, Sa\pi = 2a^2\pi^2,$

is the equation required.

Ex. 9. *If $ABCD$ is a parallelogram, and if a circle be described passing through the point A, and cutting the sides AB, AC and the diagonal AD in the points F, G, H respectively; then the rectangle $AD \cdot AH$ is equal to the sum of the rectangles $AB \cdot AF$, and $AC \cdot AG$.*

Let $\qquad AB = a$, $AC = \beta$, $AD = \gamma$

$$= a + \beta\,;$$

$$AF = xa, \quad AG = y\beta, \quad AH = z\gamma\,;$$

θ the vector diameter of the circle; then

$$xa^2 - Sa\theta = 0,$$
$$y\beta^2 - S\beta\theta = 0,$$
$$z\gamma^2 - S\gamma\theta = 0;$$

whence, since

$$\gamma = a + \beta,$$
$$z\gamma^2 = xa^2 + y\beta^2;$$

i.e. $AD . AH = AB . AF + AC . AG.$

Ex. 10. *What is represented by the equation*

$$\rho = (a + x\beta)^{-1} ?$$

If a, β be not at right angles to one another, we can put $a_1 + e\beta$ for a, and so choose e that $Sa_1\beta = 0$.

We shall therefore consider a, β as vectors at right angles to each other, and we may, on account of x, assume their tensors equal, and each a unit.

Hence

$$\rho = \frac{a + x\beta}{(a + x\beta)^2} = -\frac{a + x\beta}{1 + x^2},$$

or, if

$$\sin \theta = \frac{1}{\sqrt{1 + x^2}},$$
$$\cos \theta = -\frac{x}{\sqrt{1 + x^2}},$$
$$\rho = -\sin \theta (a \sin \theta + \beta \cos \theta),$$

whence

$$T\rho (= r) = \sin \theta,$$

a circle of which the diameter is a unit parallel to a and the origin a point in the circumference; and β a tangent vector at the origin.

Otherwise,

$$Sa\rho = \frac{1}{1 + x^2},$$
$$S\beta\rho = \frac{x}{1 + x^2};$$
$$\therefore (Sa\rho)^2 + (S\beta\rho)^2 = Sa\rho,$$
$$\text{or } -\rho^2 = Sa\rho.$$

Or, again,
$$\rho^{-1} = a + x\beta\,;$$

whence
$$Sa\rho^{-1} = -1,$$

$$\text{or}\quad V\beta\,(\rho^{-1} - a) = 0,$$

$$\text{or}\quad U\,\frac{\rho^{-1} - a}{\beta} = 1,$$

where U stands for the versor of the quaternion;

all of these being, with the obvious condition $S\,.\,a\beta\rho = 0$, varieties of the form of the equation of a, circle, referred to a point in the circumference, the diameter through which is parallel to a.

Draw any two radii ρ and ρ_1, then we have

$$S\,.\,U\rho^{-1}U\,(\rho_1^{-1} - \rho^{-1}) = S\,.\,U\rho^{-1}U\frac{\rho_1\rho^2 - \rho_1^2\rho}{\rho_1^2\rho^2}$$

$$= S\,.\,U\rho^{-1}U\frac{\rho_1\,(\rho - \rho_1)\,\rho}{\rho_1^2\rho^2}\,.$$

Now $\dfrac{\rho_1\,(\rho - \rho_1)\,\rho}{\rho_1^2\rho^2}$ will be rendered a unit if we take a unit

vector along each of the three vectors ρ_1, $(\rho - \rho_1)$, and ρ;

$$\therefore\;\; S\,.\,U\rho^{-1}U\,(\rho_1^{-1} - \rho^{-1}) = S\,.\,U\rho^{-1}U\rho_1\,U\,(\rho - \rho_1)\,U\rho$$

$$= S\,.\,U\rho_1\,U\,(\rho - \rho_1).$$

But
$$\rho_1^{-1} - \rho^{-1} = (x_1 - x)\,\beta\,;$$

$$\therefore\;\; U\,(\rho_1^{-1} - \rho^{-1}) = \beta,$$

and
$$S\,.\,U\rho^{-1}U\,(\rho_1^{-1} - \rho^{-1}) = S\beta U\rho^{-1} = -S\beta U\rho.$$

Hence
$$S\,.\,U\rho_1 U\,(\rho - \rho_1) = -S\beta U\rho.$$

If ρ be constant whilst ρ_1 varies, the right-hand side of this equation is constant, and the equation shews that the angles in the same segment of a circle are equal to one another.

Further, the form of the right-hand side of the equation, viz. $-S\beta U\rho$, shews that the angle in the segment is equal to the supplement of the angle between the chord (ρ) and the tangent (β).

38. *To draw a tangent to a circle.*

1.   If we assume the first form of the equation, the centre being the origin, and assume also that the tangent is at right

angles to the radius drawn to the point of contact; we shall have, denoting by π a vector to a point in the tangent,

$$S\rho\,(\pi-\rho)=0,$$

for $\pi-\rho$ is along the tangent;

$$\therefore\;\; S\pi\rho=-\,a^2$$

is the equation required.

2. Without assuming the property of the tangent, we may obtain it as follows.

Let ρ' be a point in the circle near to P; then

$$S\,(\rho'^2-\rho^2)=0,$$

from the equation;

i.e. $S\,(\rho'+\rho)(\rho'-\rho)=0.$

But $\rho'+\rho$ is the vector which bisects the angle between the vectors to the points of section, and $\rho'-\rho$ is a vector along the secant.

Now the equation shews (22. 7) that the former of these lines is perpendicular to the latter.

As the points of section approach one another, the tangent approaches the secant, and the bisecting line approaches the radius to the point of contact: therefore the radius to the point of contact is perpendicular to the tangent.

39. From a point without a circle two tangents are drawn to the circle, to find the equation of the chord of contact.

Let β be the vector to the given point,

$$S\pi\rho=-\,a^2$$

the equation of a tangent; then since it passes through the given point

$$S\beta\rho=-\,a^2.$$

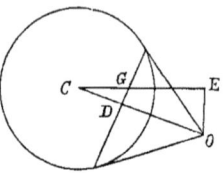

Now this equation is satisfied for both points of contact, and since it is the equation of a straight line (32. 3) it must be satisfied for every point in the straight line which passes through those points: it is therefore the equation of the chord of contact. To avoid the appearance of limiting ρ to a point in the circle, we may write σ in place of ρ; and the equation of the chord of contact becomes

$$S\beta\sigma = -a^2.$$

EXAMPLES.

40. Ex. 1. *If chords be drawn through a given point, and tangents be drawn at the points of section, the corresponding pairs of tangents will intersect in a straight line.*

Let γ be the vector to the given point G, the centre C being the origin; β the vector to O, the point of intersection of two tangents at the extremities of a chord through G; then the equation of the chord of contact is (39)

$$S\beta\sigma = -a^2,$$

and as the chord passes through G we have

$$S\beta\gamma = -a^2,$$

which, since γ is a constant vector, is the equation of a straight line, the locus of β.

COR. 1.   The straight line is at right angles to CG (32. 3).

COR. 2.   The converse is obviously true, that if through points in a straight line pairs of tangents be drawn to a circle, the chords of contact all pass through the same point.

Ex. 2.   *Any chord drawn from O the point of intersection of two tangents, is cut harmonically by the circle and the chord of contact.*

Let radius $= a$, $OC = c$, $OR = p$, $OS = q$, vector $OC = \alpha$, unit vector $OR = \rho$; then

$$(p\rho)^2 - 2pS\alpha\rho = c^2 - a^2$$

is the equation of the circle;

i.e. $p^2 + 2pS\alpha\rho + c^2 - a^2 = 0$,

6

a quadratic equation which gives the two values of p, viz. OR and OT;

$$\therefore \frac{1}{OR} + \frac{1}{OT} = -\frac{2Sa\rho}{c^2 - a^2}.$$

But $q\rho = OS = ON + NS$,

$$Saq\rho = SaON;$$

i.e. $qSa\rho = Sa\,(OC - NC)$

$$= a^2 - SaNC$$

$$= -c^2 + a^2 \ (39);$$

hence

$$\frac{2}{OS} = \frac{2}{q}$$

$$= -\frac{2Sa\rho}{c^2 - a^2}$$

$$= \frac{1}{OR} + \frac{1}{OT}.$$

Ex. 3.　*If tangents be drawn at the angular points of a triangle inscribed in a circle, the intersections of these tangents with the opposite sides of the triangle lie in a straight line.*

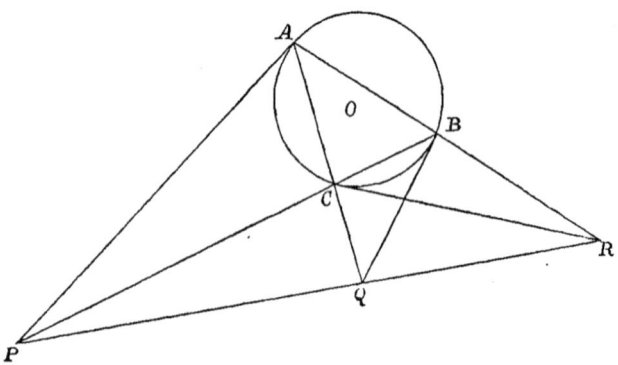

Let radius $= a$, $OA = a$, $OB = \beta$, $OC = \gamma$, then

$$OP = a + xAP = \beta + y\,(\gamma - \beta).$$

But a is perpendicular to AP;

$$\therefore\ Sa^2 = Sa\beta + yS\,(a\gamma - a\beta),$$

$$y = -\frac{a^2 + Sa\beta}{Sa\gamma - Sa\beta},$$

and

$$OP = \frac{(a^2 + Sa\gamma)\,\beta - (a^2 + Sa\beta)\,\gamma}{Sa\gamma - Sa\beta}.$$

Similarly,

$$OQ = \frac{(a^2 + Sa\beta)\,\gamma - (a^2 + S\beta\gamma)\,a}{Sa\beta - S\beta\gamma},$$

$$OR = \frac{(a^2 + S\beta\gamma)\,a - (a^2 + Sa\gamma)\,\beta}{S\beta\gamma - Sa\gamma}.$$

Hence

$$(Sa\gamma - Sa\beta)\,OP + (Sa\beta - S\beta\gamma)\,OQ + (S\beta\gamma - Sa\gamma)\,OR = 0,$$

whilst $(Sa\gamma - Sa\beta) + (Sa\beta - S\beta\gamma) + (S\beta\gamma - Sa\gamma) = 0.$

Consequently (Art. 13) P, Q, R are in the same straight line.

COR. $PQ\ :\ PR\ ::\ S\beta\gamma - Sa\gamma\ ::\ S\beta\gamma - Sa\beta$

$$::\ \cos 2B - \cos 2A\ :\ \cos 2C - \cos 2A$$

$$::\ \sin C \sin (B - A)\ :\ \sin B \sin (C - A).$$

Ex. 4.   *A fixed circle is cut by a number of circles, all of which pass through two given points; to prove that the lines of section of the fixed circle with each circle of the series all pass through a point whose distances from the two given points are proportional to the squares of the tangents drawn from those points to the fixed circle.*

Let O be the centre of the fixed circle whose radius is a, A, B the given points, vectors a, β, the origin being O; $OA = b$, $OB = c$; C the centre of a circle which passes through A and B, radius r; $OC = \rho$, π the vector to any point in the circumference of this circle; then the equation of the circle is $(\pi - \rho)^2 = -r^2$;

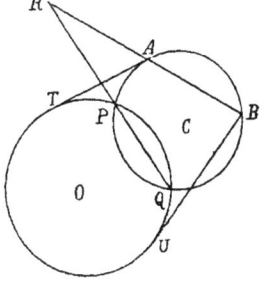

hence for the four points A, B, P, Q, we have

$$a^2 - 2Sa\rho + \rho^2 = -r^2,$$
$$\beta^2 - 2S\beta\rho + \rho^2 = -r^2,$$
$$OP^2 - 2S \cdot OP\rho + \rho^2 = -r^2,$$
$$OQ^2 - 2S \cdot OQ\rho + \rho^2 = -r^2.$$

From which it follows that

$$S(OP - OQ)\rho = 0 \dots\dots\dots\dots\dots(1),$$
$$-b^2 + c^2 = a^2 - \beta^2 = 2S(a - \beta)\rho \dots\dots\dots\dots(2),$$
$$2S(OP - a)\rho = OP^2 - a^2 = -a^2 + b^2 \dots\dots\dots\dots(3).$$

Let QP, AB intersect in R, $OR = \sigma$; then

$$S\sigma\rho = S\{OP + x(OP - OQ)\}\rho$$
$$= S \cdot OP\rho \text{ by } (1),$$

and
$$S\sigma\rho = S\{a + y(a - \beta)\}\rho$$
$$= Sa\rho + \frac{y}{2}(-b^2 + c^2) \text{ by } (2);$$

$$\therefore y(-b^2 + c^2) = 2S\sigma\rho - 2Sa\rho$$
$$= 2S(OP - a)\rho$$
$$= -a^2 + b^2 \text{ by } (3),$$

i.e. y is independent of ρ and r; or R is the same point for every circle:

also
$$OR = \frac{(c^2 - a^2)a - (b^2 - a^2)\beta}{c^2 - b^2},$$

and $RA : RB :: a - OR : \beta - OR :: b^2 - a^2 : c^2 - a^2$
$$:: AT^2 : BU^2.$$

41. *The Sphere.*

1. It is clear that there is nothing in the demonstration of Art. 36 which limits the conclusions to one plane; it follows that the equations there obtained are also equations of a sphere.

2. Further if we assume that the tangent plane to a sphere is perpendicular to the radius to the point of contact, the conclusion in Art. 38 is applicable also.

The equation of the tangent plane to the sphere is therefore
$$S\pi\rho = -a^2.$$

3. Lastly, the results of Art. 39 are also applicable if we substitute any number of tangent planes passing through a given point for two tangent lines; the equation of the plane which passes through the points of contact is therefore
$$S\beta\sigma = -a^2.$$

This plane is the *polar* plane to the point through which the tangent planes pass.

Cor. Since the polar plane is perpendicular to the line which joins the centre with the point through which the tangent planes pass, the perpendicular CD to it from the centre is along this line and has therefore the same unit vector with it. The equation above gives in this case
$$S\{CO \cdot CD \,(U\beta)^2\} = -a^2 ;$$
$$\therefore CO \cdot CD = a^2 \quad (19).$$

Examples.

42. Ex. 1. *Every section of a sphere made by a plane is a circle.*

Let $\rho^2 = -a^2$ be the equation of the sphere, α the vector perpendicular from the centre on the cutting plane; c the corresponding line.

Let $$\rho = \alpha + \pi ;$$
then the equation becomes
$$\pi^2 + 2S\alpha\pi + \alpha^2 = -a^2.$$
But $$S\alpha\pi = 0 ;$$
$$\therefore \pi^2 = -(a^2 - c^2)$$
is the equation of the section, which is therefore a circle, the square of whose radius is $a^2 - c^2$.

Ex. 2. *To find the curve of intersection of two spheres.*

Let the equations be
$$\rho^2 - 2S\alpha\rho = C,$$
$$\rho^2 - 2S\alpha'\rho = C' ;$$

$$\therefore\ 2S\left(a' - a\right)\rho = C - C',$$

a plane perpendicular to the line of which the vector is $a' - a$, i.e. to the line which joins the centres of the two spheres.

Hence, by Ex. 1, the curve of intersection is a circle.

Ex. 3. *To find the locus of the feet of perpendiculars from the origin on planes which pass through a given point.*

Let a be the vector to the point, δ perpendicular on a plane through it; then

$$S\delta\left(\rho - a\right) = 0$$

is the equation of that plane; therefore for the foot of the perpendicular

$$S\left(\delta^2 - a\delta\right) = 0\ ;$$

or $$\delta^2 - Sa\delta = 0$$

is true for the foot of every perpendicular and is therefore the equation of the surface required. Hence it is a sphere whose diameter is the line joining the origin with the given point.

Ex. 4. *Perpendiculars are drawn from a point on the surface of a sphere to all tangent planes, to find the locus of their extremities.*

Let a be the vector to the given point,

$$S\pi\rho = -a^2$$

the equation of a tangent plane.

Since the perpendicular is parallel to ρ, its vector is

$$\pi = a + x\rho\ ;$$
$$\therefore\ \left(\pi - a\right)^2 = x^2\rho^2 = x^2a^2$$
$$= -x^2a^2\ ;$$

because both ρ and a are vector radii.

But $S\pi\rho = -a^2$ gives with $x\rho = \pi - a$,

$$S\pi\left(\pi - a\right) = -a^2x,$$
$$\left(\pi^2 - Sa\pi\right)^2 = a^4x^2$$
$$= -a^2 \times -a^2x^2$$
$$= -a^2\left(\pi - a\right)^2.$$

Ex. 5. *If the points from which tangent planes are drawn to a sphere lie always in a straight line, prove that the planes of section all pass through a given point.*

Let CE be perpendicular to the line in which the point β lies (41), see fig. of Art. 39,

$$CE = c, \ \text{vector} \ CE = \delta;$$

then $$S\beta\delta = -c^2$$

is the equation of the line.

But $$S\beta\sigma = -a^2$$

is the plane of contact, which is therefore satisfied by

$$\sigma = \frac{a^2}{c^2}\delta,$$

i.e. the planes all pass through a point G in CE, such that

$$CG = \frac{a^2}{c^2} CE,$$

$$\text{or} \ \ CE \cdot CG = a^2.$$

Ex. 6. *If three spheres intersect one another, their three planes of intersection all pass through the same straight line.*

Let a, β, γ be the vectors to the centres of the three spheres,

$$\rho^2 - 2Sa\rho = a,$$
$$\rho^2 - 2S\beta\rho = b,$$
$$\rho^2 - 2S\gamma\rho = c,$$

their three equations ;

$$\therefore \ 2S(a - \beta)\rho = b - a,$$
$$2S(a - \gamma)\rho = c - a,$$
$$2S(\beta - \gamma)\rho = c - b,$$

are the equations of the three planes of intersection.

Now the line of intersection of the first and second of these planes is obtained by taking ρ so as to satisfy both equations, and therefore their difference

$$2S(\beta - \gamma)\rho = c - b,$$

which, being the third equation, proves that the same value of ρ satisfies it also. The three planes consequently all pass through the same straight line.

Ex. 7. *To find the locus of a point, the sum of the squares of whose distances from a number of given points has a given value.*

Let ρ denote the sought point; $a, \beta, \ldots$ the given ones; then

$$(\rho - a)^2 + (\rho - \beta)^2 + \&c. = \Sigma (\rho - a)^2 = - C.$$

If there be n given points; this is

$$n\rho^2 - 2S . \rho\Sigma a + \Sigma a^2 = - C,$$

or

$$\left(\rho - \frac{\Sigma a}{n}\right)^2 = \left(\frac{\Sigma a}{n}\right)^2 - \frac{1}{n}(\Sigma . a^2 + C).$$

This is the equation of a sphere, the vector to whose centre is

$$\frac{1}{n} \Sigma (a),$$

i.e. the centre of inertia of the n points taken as equal.

Transpose the origin to this point, then (36)

$$\Sigma . a = 0,$$

and

$$\rho^2 = - \frac{1}{n} \{\Sigma (a^2) + C\}.$$

Hence, that there may be a real locus, C must be positive and not less than the sum of the squares of the distances of the given system of points from their centre of inertia. If C have its least value, we have of course

$$\rho^2 = 0,$$

the sphere having shrunk to a point.

ADDITIONAL EXAMPLES TO CHAP. V.

1. If two circles cut one another, and from one of the points of section diameters be drawn to both circles, their other extremities and the other point of section will be in a straight line.

2. If a chord be drawn parallel to the diameter of a circle, the radii to the points where it meets the circle make equal angles with the diameter.

3. The locus of a point from which two unequal circles subtend equal angles is a circle.

4. A line moves so that the sum of the perpendiculars on it from two given points in its plane is constant. Shew that the locus of the middle point between the feet of the perpendiculars is a circle.

5. If O, O' be the centres of two circles, the circumference of the latter of which passes through O; then the point of intersection A of the circles being joined with O' and produced to meet the circles in C, D, we shall have

$$AC . AD = 2AO^2.$$

6. If two circles touch one another in O, and two common chords be drawn through O at right angles to one another, the sum of their squares is equal to the square of the sum of the diameters of the circles.

7. A, B, C are three points in the circumference of a circle; prove that if tangents at B and C meet in D, those at C and A in E, and those at A and B in F; then AD, BE, CF will meet in a point.

8. If A, B, C are three points in the circumference of a circle, prove that $V(AB . BC . CA)$ is a vector parallel to the tangent at A.

9. A straight line is drawn from a given point O to a point P on a given sphere : a point Q is taken in OP so that

$$OP . OQ = k^2.$$

Prove that the locus of Q is a sphere.

10. A point moves so that the ratio of its distances from two given points is constant. Prove that its locus is either a plane or a sphere.

11.   A point moves so that the sum of the squares of its distances from a number of given points is constant.   Prove that its locus is a sphere.

12.   A sphere touches each of two given straight lines which do not meet; find the locus of its centre.

CHAPTER VI.

THE ELLIPSE.

43. 1. If we define a conic section as "the locus of a point which moves so that its distance from a fixed point bears a constant ratio to its distance from a fixed straight line" (Todhunter, Art. 123), we shall find the equation to be (Ex. 5, Art. 35)

$$a^2\rho^2 = e^2 (a^2 - Sa\rho)^2 \dots\dots\dots\dots\dots\dots(1),$$

where $\quad SP = ePQ$, vector $SD = a$, $SP = \rho$.

When e is less than 1, the curve is the ellipse, a few of whose properties we are about to exhibit.

2. SA, SA' are multiples of a: call one of them xa: then, by equation (1), putting xa for ρ, we get

$$x^2 = e^2 (1-x)^2 ;$$

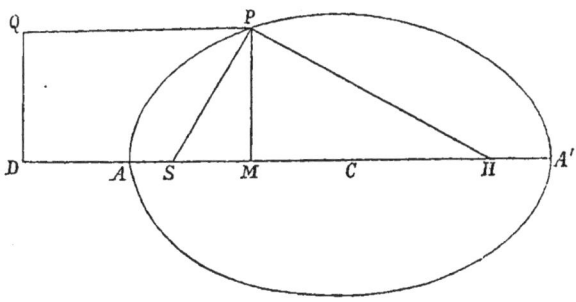

$$\therefore \ x = \frac{e}{1+e},$$

$$x = -\frac{e}{1-e},$$

i.e. $SA = \dfrac{e}{1+e} SD,$

$$SA' = \dfrac{e}{1-e} SD \, ;$$

$$\therefore \ AA' = \dfrac{2e}{1-e^2} SD,$$

the major axis of the ellipse, which we shall as usual abbreviate by $2a$.

If C be the centre of the ellipse

$$CS = SA' - CA' = \left(\dfrac{e}{1-e} - \dfrac{e}{1-e^2} \right) SD = eCA$$

$$= ae,$$

and if vector CS be designated by a', CP by ρ', we have

$$a' = \dfrac{e^2}{1-e^2} a \ \text{ and } \ \rho' = \rho + a' \, ;$$

whence, by substituting in (1), the equation assumes the form

$$a^2 \rho'^2 + (Sa'\rho')^2 = -a^4 (1 - e^2) \, ;$$

which we may now write, CS being α and CP ρ,

$$a^2\rho^2 + (Sa\rho)^2 = -a^4 (1 - e^2) \dotfill (2).$$

3.   This equation might have been obtained at once by referring the ellipse to the two foci, as Newton does in the Principia, Book I. Prop. 11; the definition then becomes

$$SP + HP = 2a,$$

or in vectors, if

$$CP = \rho, \ \ CS = a,$$

$$T(\rho + a) + T(\rho - a) = 2a \, ;$$

i.e. $\sqrt{-(\rho + a)^2} + \sqrt{-(\rho - a)^2} = 2a \, ;$

hence, squaring,

$$a\sqrt{-(\rho - a)^2} = a^2 + Sa\rho \, ;$$

i.e. $a^2\rho^2 + (Sa\rho)^2 = -a^2 (1 - e^2).$

If now we write $\phi\rho$ for $-\dfrac{a^2\rho + aSa\rho}{a^4(1-e^2)}$, where $\phi\rho$ is a vector which coincides with ρ only in the cases in which either a coincides with ρ or when $Sa\rho = 0$, i.e in the cases of the principal axes; the equation of the ellipse becomes

$$S\rho\phi\rho = 1\dots\dots\dots\dots\dots\dots\dots\dots(3).$$

The same equation is, of course, applicable to the hyperbola, e being greater than 1.

44. The following properties of $\phi\rho$ will be *very frequently* employed. The reader is requested to bear them constantly in mind.

1. $\phi(\rho + \sigma) = \phi\rho + \phi\sigma$.

2. $\phi x\rho = x\phi\rho$.

3. $S\sigma\phi\rho = -\dfrac{a^2 S\sigma\rho + Sa\sigma Sa\rho}{a^4(1-e^2)}$

    $= S\rho\phi\sigma$.

They need no other demonstration than what results from simple inspection of the value of $\phi\rho$

$$= -\frac{a^2\rho + aSa\rho}{a^4(1-e^2)}.$$

45. To find the equation of the tangent to the ellipse.

The tangent is defined to be the limit to which the secant approaches as the points of section approach each other.

Let $CP = \rho$, $CQ = \rho'$, then

    vector $PQ = CQ - CP = \rho' - \rho = \beta$ say ;

β is therefore a vector along the secant.

Now $S\rho'\phi\rho' = S(\rho + \beta)\phi(\rho + \beta)$

    $= S(\rho + \beta)(\phi\rho + \phi\beta)$ (44. 1)

    $= S\rho\phi\rho + S\rho\phi\beta + S\beta\phi\rho + S\beta\phi\beta$.

But
$$S\rho'\phi\rho' = 1 = S\phi\phi\rho ;$$
$$\therefore\ S\rho\phi\beta + S\beta\phi\rho + S\beta\phi\beta = 0,$$
or (44. 3)
$$2S\beta\phi\rho + S\beta\phi\beta = 0.$$

Now $\beta\phi\rho$ involves the first power of β whilst $\beta\phi\beta$ involves the second, and the definition requires that the limit of the sum of the two as β gets smaller and smaller should be the first only, even if that should be zero : i.e. when β is along the tangent, we must have
$$2S\beta\phi\rho = 0.$$

[We might also have written the equation in the form
$$2S.\ \beta\left(\phi\rho + \frac{1}{2}\phi\beta\right) = 0.$$

Thus, however small the tensor of β may be,
$$\phi\left(\rho + \frac{1}{2}\beta\right)$$

is always perpendicular to β. Whence, finally,
$$S\beta\phi\rho = 0.]$$

Let then T be any point in the tangent, vector $CT = \pi$, then
$$\pi = \rho + x\beta,$$
and $S\beta\phi\rho = 0$ gives
$$S\left(\pi - \rho\right)\phi\rho = 0 ;$$
$$\therefore\ S\pi\phi\rho = S\rho\phi\rho = 1$$

is the equation of the tangent.

Cor. 1 $\phi\rho$ is a vector along the perpendicular to the tangent (32. 3), that is, $\phi\rho$ is a normal vector, or parallel to a normal vector at the point ρ.

Cor. 2. The equation of the tangent may also be written (44. 3) $S\rho\phi\pi = 1$.

46. We may now exhibit the corresponding equations in terms of the Cartesian co-ordinates, as some of the results are best known in that form.

Let $CM = x$, $MP = y$ as usual; then, retaining the notation of Art. 31 with i, j as unit vectors parallel and perpendicular respectively to CA,

$$\text{vector } CM = xi, \quad MP = yj, \quad CS = aei;$$

$$\therefore \rho = xi + yj,$$

$$\phi\rho = -\frac{a^2\rho + aSa\rho}{a^4(1 - e^2)}$$

$$= -\frac{a^2(1 - e^2)xi + a^2 yj}{a^4(1 - e^2)}$$

$$= -\left(\frac{xi}{a^2} + \frac{yj}{b^2}\right),$$

where $\qquad b^2 = a^2(1 - e^2)$;

and $\qquad S\rho\phi\rho = -S(xi + yj)\left(\frac{xi}{a^2} + \frac{yj}{b^2}\right)$

$$= \frac{x^2}{a^2} + \frac{y^2}{b^2};$$

$$\therefore \frac{x^2}{a^2} + \frac{y^2}{b^2} = 1$$

is the Cartesian interpretation of $S\rho\phi\rho = 1$.

Again, if x', y' be the co-ordinates of T a point in the tangent,

$$\pi = x'i + y'j,$$

and $\qquad S\pi\phi\rho = -S(x'i + y'j)\left(\frac{xi}{a^2} + \frac{yj}{b^2}\right)$

$$= \frac{xx'}{a^2} + \frac{yy'}{b^2};$$

$$\therefore \frac{xx'}{a^2} + \frac{yy'}{b^2} = 1$$

is the equation of the tangent.

47. The values of ρ and $\phi\rho$ exhibited in the last Article, viz.

$$\rho = xi + yj, \quad \phi\rho = -\left(\frac{xi}{a^2} + \frac{yj}{b^2}\right)\dots\dots\dots\dots\dots(1),$$

enable us to write

$$\phi\rho = \frac{iSi\rho}{a^2} + \frac{jSj\rho}{b^2}\dots\dots\dots\dots\dots\dots(2).$$

We shall have

$$\phi^2\rho = \phi\phi\rho = \frac{iSi\phi\rho}{a^2} + \frac{jSj\phi\rho}{b^2}$$

$$= -\left(\frac{iSi\rho}{a^4} + \frac{jSj\rho}{b^4}\right)\dots\dots\dots\dots(3),$$

$$\phi^{-1}\rho = a^2 iSi\rho + b^2 jSj\rho, \text{ \&c.}$$

If, further, we write $\psi\rho$ for

$$-\left(\frac{iSi\rho}{a} + \frac{jSj\rho}{b}\right),$$

we shall have

$$\psi^2\rho = \psi\psi\rho = -\left(\frac{iSi\rho}{a^2} + \frac{jSj\rho}{b^2}\right)$$

$$= -\phi\rho\dots\dots\dots\dots\dots\dots\dots\dots(4),$$

$$\psi^{-1}\rho = -(aiSi\rho + bjSj\rho), \text{ \&c.}$$

$$\rho = \psi^{-1}\psi\rho$$

$$= -(aiSi\psi\rho + bjSj\psi\rho)\dots\dots\dots\dots\dots(5).$$

It is evident that the properties of $\phi\rho$ (Art. 44) are possessed by all these functions.

Now $\qquad\qquad S\rho\phi\rho = 1$

gives $\qquad\qquad S\rho\psi(\psi\rho) = -1.$

But since $\qquad\qquad S\rho\psi\sigma = S\sigma\psi\rho,$

this becomes $\qquad S\psi\rho\psi\rho = -1,$

or $\qquad\qquad\qquad T\psi\rho = 1;$

which shews 1. that $\psi\rho$ is a unit vector ; 2. that the equation of the ellipse may be expressed in the *form* of the equation of a circle, the vector which represents the radius being itself of variable length, deformed by the function ψ.

Lastly,　　　　　　　　　　$Sa\phi\beta = 0$

gives　　　　　　　　$Sa\psi^2\beta = S\psi a\psi\beta = 0$;

therefore $\psi a,\ \psi\beta$ are vectors at right angles to one another.

48. To find the locus of the middle points of parallel chords.

Let all the chords be parallel to the vector β ; π the vector to the middle point of one of them whose vector length is $2x\beta$; then

$$\pi + x\beta, \quad \pi - x\beta$$

are vectors to points in the ellipse ;

$$\therefore\ S\left(\pi + x\beta\right)\phi\left(\pi + x\beta\right) = 1,$$
$$S\left(\pi - x\beta\right)\phi\left(\pi - x\beta\right) = 1,$$

multiplying out, observing that (44. 1),

$$\phi\left(\pi + x\beta\right) = \phi\pi + x\phi\beta,\ \&\text{c.,}$$

we get by subtracting,

$$S\pi\phi\beta + S\beta\phi\pi = 0,$$

or, (Art. 44. 3),

$$2S\pi\phi\beta = 0\ ;$$
$$\therefore\ S\pi\phi\beta = 0,$$

i. e. the locus required is a straight line perpendicular to $\phi\beta$.

Now $\phi\beta$ is the vector perpendicular to the tangent at the extremity of the diameter β (Art. 45. Cor. 1).

Therefore the locus of the middle points of parallel chords is a diameter parallel to the tangent at the extremity of the diameter to which the chords are parallel.

Cor. If a be the diameter which bisects all chords parallel to β ; since

$$Sa\phi\beta = 0,.$$

we have (Art. 44. 3),

$$S\beta\phi\alpha = 0,$$

which is the equation to the straight line that bisects all chords parallel to α. Moreover β is parallel to the tangent at the extremity of α, for it is perpendicular to the normal $\phi\alpha$.

Hence the properties of α with respect to β are convertible with those of β with respect to α: and the diameters which satisfy the equation

$$S\alpha\phi\beta = 0,$$

are said to be conjugate to one another.

49. Our object being simply to illustrate the process, we shall set down in this Article a few of the properties of conjugate diameters without attempting to classify or complete them.

1. If CP, CD are the conjugate semi-diameters α, β; and if DC be produced to meet the ellipse again in E, and PD, PE be joined; vector $DP = \alpha - \beta$, vector $EP = \alpha + \beta$.

Now

$$\begin{aligned}
S\,(\alpha + \beta)\,\phi\,(\alpha - \beta) &= S\,(\alpha + \beta)\,(\phi\alpha - \phi\beta) \\
&= S\alpha\phi\alpha - S\beta\phi\beta - S\alpha\phi\beta + S\beta\phi\alpha \quad (44.\ 1) \\
&= 0,
\end{aligned}$$

because $S\alpha\phi\alpha$, $S\beta\phi\beta$, each equals 1.

Therefore $\alpha + \beta$, $\alpha - \beta$ are parallel to conjugate diameters. (Art. 48. Cor.)

This is the property of *Supplemental Chords*.

2. Let two tangents meet in T, $CT = \pi$, and let the chord of contact be parallel to β. If for the present purpose we denote CN by α, we have

$$S\pi\phi\,(\alpha + x\ \beta) = 1,$$

$$S\pi\phi\,(\alpha + x_{,}\ \beta) = 1,$$

for the two points of contact.

Subtracting and applying (44. 1),

$$S\pi\phi\beta = 0 :$$

hence π and β i.e. CT, QR are conjugate.

3.   The equation of the chord of contact is $S\sigma\phi\pi = 1$.

For $S\rho\phi\pi = 1$ (45. Cor. 2) is satisfied by the values of ρ at Q and at R, and since $S\rho\phi\pi = 1$ or $S\sigma\phi\pi = 1$ is the equation

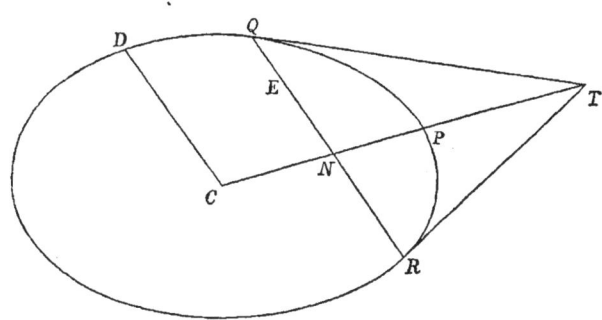

of a straight line, π being a constant vector (32. 3) it is the line QR.

4.   If QR pass through a fixed point E, the locus of T is a straight line.

Let σ be the vector to the point E, then

$$S\sigma\phi\pi = 1 ;$$
$$\therefore \ \ S\pi\phi\sigma = 1,$$

or the locus of T is a straight line perpendicular to $\phi\sigma$, i.e. parallel to the tangent at the point where CE meets the ellipse. (45. Cor. 1.)

The converse is of course true.

5.   Let us now take

$$CP = \alpha, \ \ CD = \beta, \ \ CN = x\alpha, \ \ NQ = y\beta, \ \ CT = z\alpha ;$$

then the equation of the tangent becomes

$$Sza\phi\,(xa + y\beta) = 1\ ;$$

$$\text{i.e. } xzSa\phi a = 1\ ;$$

$$\therefore\ xz = 1,$$

$$\text{or } xa\,.\,za = a^2\ ;$$

geometrically $\qquad CN\,.\,CT = CP^2.$

6. The equation of the ellipse gives

$$S\,(xa + y\beta)\,\phi\,(xa + y\beta) = 1,$$

or $\qquad x^2 Sa\phi a + y^2 S\beta\phi\beta + 2xy Sa\phi\beta = 1,$

$$\text{i.e. } x^2 + y^2 = 1,$$

or, since CN is xa, $CP = a$, &c.,

$$\left(\frac{CN}{CP}\right)^2 + \left(\frac{NQ}{CD}\right)^2 = 1,$$

the equation of the ellipse referred to conjugate diameters.

7. $\qquad a = \psi^{-1}\psi a = -\,(ai Si\psi a + bj Sj\psi a)$

$$\beta = \psi^{-1}\psi\beta = -\,(ai Si\psi\beta + bj Sj\psi\beta)\ ;$$

$$\therefore\ Va\beta = ab\,Vij\,(Si\psi a Sj\psi\beta - Si\psi\beta Sj\psi a).$$

If now we call k the unit vector perpendicular to the plane of the ellipse, we get

$$Vij = k.$$

And, observing that ψa, $\psi\beta$ are unit vectors at right angles; if the angle between i and ψa be θ, that between i and $\psi\beta$ will be

$$\frac{\pi}{2} + \theta,\ \text{&c. &c.,}$$

we shall have (21. 3)

$$Si\psi a = -\cos\theta,$$

$$Si\psi\beta = \sin\theta,$$

$$Sj\psi a = -\sin\theta,$$

$$Sj\psi\beta = -\cos\theta.$$

$$\therefore\ Si\psi a Sj\psi\beta - Si\psi\beta Sj\psi a = \cos^2\theta + \sin^2\theta = 1.$$

Consequently $\qquad V\alpha\beta = abk$;

i.e. $\quad T\alpha \cdot T\beta \sin PCD = ab$,

or *all parallelograms circumscribing an ellipse are equal.*

50. Examples.

Ex. 1. *To find the length of the perpendicular from the centre on the tangent.*

Let CY the perpendicular, which (Art. 45. Cor. 1) is a vector along $\phi\rho$, be $x\phi\rho$; then since Y is a point in the tangent,

$$S\pi\phi\rho = 1 \quad \text{gives} \quad Sx\phi\rho\phi\rho = 1, \cdot$$

$$\text{or} \quad x(\phi\rho)^2 = 1 ;$$

$$\therefore \quad (x\phi\rho)^2(\phi\rho)^2 = 1,$$

and $\qquad CY^2 = T(x\phi\rho)^2 = T\dfrac{1}{(\phi\rho)^2}$

$$= \dfrac{1}{\dfrac{x^2}{a^4} + \dfrac{y^2}{b^4}} \quad (46).$$

Ex. 2. *The product of the perpendiculars from the foci on the tangent is equal to the square of the semi-axis minor.*

We have SY the vector perpendicular $= x\phi\rho$, and as Y is a point in the tangent, and

$$CY = \alpha + x\phi\rho,$$

$$S(\alpha + x\phi\rho)\phi\rho = 1,$$

$$x(\phi\rho)^2 = 1 - S\alpha\phi\rho,$$

$$SY = Tx\phi\rho = T\dfrac{1 - S\alpha\phi\rho}{\phi\rho} .$$

Similarly, $\qquad HZ = T\dfrac{1 + S\alpha\phi\rho}{\phi\rho} ;$

$$\therefore \quad SY \cdot HZ = T\dfrac{1 - S^2\alpha\phi\rho}{(\phi\rho)^2} .$$

Now (43. 2) $a^2\rho^2 = -S^2a\rho - a^4(1-e^2)$,

$$\phi\rho = -\frac{a^2\rho + aSa\rho}{a^4(1-e^2)} \; ;$$

$$\therefore (\phi\rho)^2 = \frac{S^2a\rho - a^4}{a^6(1-e^2)} \; ,$$

$$1 - S^2(a\phi\rho) = \frac{a^4 - S^2a\rho}{a^4} .$$

$$\therefore SY.HZ = a^2(1-e^2) = b^2.$$

Ex. 3. *The perpendicular from the focus on the tangent intersects the tangent in the circumference of the circle described about the axis major.*

Retaining the notation of the last example, we have

$$CY = a + x\phi\rho$$

$$= a + \frac{\phi\rho(1-Sa\phi\rho)}{(\phi\rho)^2} \; ;$$

$$\therefore CY^2 = a^2 + \frac{2Sa\phi\rho(1-Sa\phi\rho)}{(\phi\rho)^2} + \frac{(1-Sa\phi\rho)^2}{(\phi\rho)^2}$$

$$= a^2 + \frac{1-S^2a\phi\rho}{(\phi\rho)^2}$$

$$= -a^2e^2 - a^2(1-e^2) \text{ (last example)}$$

$$= -a^2,$$

and the line $CY = a$.

Ex. 4. *To find the locus of T when the perpendicular from the centre on the chord of contact is constant.*

If CT be π, the equation of QR, the chord of contact, is

$$S\sigma\phi\pi = 1 \quad \text{(Art. 49. 3)},$$

and the perpendicular (Ex. 1) is $T\dfrac{1}{\phi\pi}$;

$$\therefore \; (\phi\pi)^2 = -c^2,$$

$$\text{or} \quad S\phi\pi \cdot \phi\pi = -c^2,$$

$$\text{or} \quad S\pi\phi\phi\pi = -c^2 \; (\text{Art. 44. 3});$$

$$\text{i.e.} \quad S\pi \left(\frac{iSi\pi}{a^4} + \frac{jSj\pi}{b^4} \right) = c^2 \; (47.\;3),$$

$$\text{or} \quad \frac{x^2}{a^4} + \frac{y^2}{b^4} = c^2,$$

an ellipse.

Ex. 5.　*TQ, TR are two tangents to an ellipse, and CQ', CR'
are drawn to the ellipse parallel respectively to TQ, TR; prove
that Q'R' is parallel to QR.*

Let　　　　　　　　$CQ = \rho, \quad CR = \rho', \quad CT = a,$

then　　　　　　　　　　$S\rho\phi a = 1,$

$$S\rho'\phi a = 1.$$

Now since CQ' is parallel to TQ,

$$CQ' = xTQ = x\,(\rho - a).$$

Similarly　　　　　　$CR' = y\,(\rho' - a),$

and　　　　　　　　$S \cdot CQ'\phi\,(CQ') = 1$

gives　　　　　　$x^2 S\,(\rho - a)\,\phi\,(\rho - a) = 1,$

i.e. $x^2\,(Sa\phi a - 1) = 1,$

and　　　　　　　　$y^2\,(Sa\phi a - 1) = 1;$

$$\therefore \; y = x,$$

and　　　　　　$Q'R' = CR' - CQ' = x\,(\rho' - \rho)$

$$= xQR;$$

hence $Q'R'$ is parallel to QR.

Cor.　　$Q'R'^2 : QR^2 :: x^2 : 1$

$$:: 1 : Sa\phi a - 1$$

$$:: 1 : \frac{x^2}{a^2} + \frac{y^2}{b^2} - 1,$$

where x, y are the co-ordinates of T.

Ex. 6.   *If a parallelogram be inscribed in an ellipse, its sides are parallel to conjugate diameters.*

Let $PQRS$ be the parallelogram.

$$PQ = a, \quad PS = \beta,$$
$$CP = \rho, \quad CS = \rho';$$

then
$$CQ = \rho + a, \quad CR = \rho' + a;$$
$$\therefore \ S\rho\phi\rho = 1,$$
$$S(\rho + a)\phi(\rho + a) = 1;$$

wherefore
$$2S\rho\phi a + Sa\phi a = 0.$$

Similarly
$$2S\rho'\phi a + Sa\phi a = 0;$$
$$\therefore \ S(\rho' - \rho)\phi a = 0, \text{ by subtraction,}$$
$$\text{or } \ S\beta\phi a = 0,$$

and (48. Cor.) β, a are parallel to conjugate diameters.

ADDITIONAL EXAMPLES TO CHAP. VI.

1.   Shew that the locus of the points of bisection of chords to an ellipse, all of which pass through a given point, is an ellipse.

2.   The locus of the middle points of all straight lines of constant length terminated by two fixed straight lines, is an ellipse whose centre bisects the shortest distance between the fixed lines; and whose axes are equally inclined to them.

3.   If chords to an ellipse intersect one another in a given point, the rectangles by their segments are to one another as the squares of semi-diameters parallel to them.

4.   If PCP', DCD' are conjugate diameters, then PD, PD' are proportional to the diameters parallel to them.

5.   If Q be a point in the focal distance SP of an ellipse, such that SQ is to SP in a constant ratio, the locus of Q is a similar ellipse.

6.   Diameters which coincide with the diagonals of the parallelogram on the axes are equal and conjugate.

7.   Also diameters which coincide with the diagonals of any parallelogram formed by tangents at the extremities of conjugate diameters are conjugate.

8.   The angular points of these parallelograms lie on an ellipse similar to the given ellipse and of twice its area.

9.   If from the extremities of the axes of an ellipse four parallel lines be drawn, the points in which they cut the curve are the extremities of conjugate diameters.

10.   If from the extremity of each of two semi-diameters ordinates be drawn to the other, the two triangles so formed will be equal in area.

11.   Also if tangents be drawn from the extremity of each to meet the other produced, the two triangles so formed will be equal in area.

12.   If on the semi-axes a parallelogram be described, and about it an ellipse similar and similarly situated to the given ellipse be constructed, any chord PQR of the larger ellipse, drawn from the further extremity of the diameter CD of the smaller ellipse, is bisected by the smaller ellipse at Q.

13.   If TP, TQ be tangents to an ellipse, and PCP' be the diameter through P, then $P'Q$ is parallel to CT.

CHAPTER VII.

THE PARABOLA AND HYPERBOLA.

51. As already stated, most of the properties of the hyperbola are the same as the corresponding properties of the ellipse, and proved by the same process, e being greater than 1. There are, however, some properties both of it and of the parabola which may be conveniently developed by a process more analogous to that of the Cartesian geometry. This process we shall develope presently. In the meantime we proceed to give a brief outline of the application to the parabola of the method employed in the preceding Chapter for the ellipse.

52. If S be the focus of a parabola, DQ the directrix, we have $SP = PQ$, $SA = AD = a$.

If $SP = \rho$, $SD = a$, we have (Ex. 5, Art. 35)

$$a^2\rho^2 = (a^2 - Sa\rho)^2 \dots (1).$$

If $\phi\rho = \dfrac{\rho - a^{-1}Sa\rho}{a^2} \dots (2),$

to which the properties of $\phi\rho$ in Art. 44 evidently apply, the equation becomes

$$S\rho\,(\phi\rho + 2a^{-1}) = 1 \dots (3).$$

If ρ' be another point in the parabola, $\rho' - \rho = \beta$, the limit to which β approaches is a vector along the tangent; so that if $x\beta = \pi - \rho$, π is the vector to a point in the tangent; this gives

$$S\,(\pi - \rho)(\phi\rho + a^{-1}) = 0 \dots (4);$$

hence the equation of the tangent becomes

$$S\pi (\phi\rho + a^{-1}) + Sa^{-1}\rho = 1 \dots\dots\dots\dots\dots(5).$$

From (2) it is evident that

$$Sa\phi\rho = 0 \dots\dots\dots\dots\dots\dots\dots\dots(6),$$

so that $\phi\rho$ is a vector perpendicular to the axis.

From the same equation

$$S\rho\phi\rho = \frac{\rho^2 - a^{-2}S^2a\rho}{a^2}$$

$$= \frac{(\rho - a^{-1}Sa\rho)^2}{a^2}$$

$$= a^2 (\phi\rho)^2 \dots\dots\dots\dots\dots\dots\dots\dots(7).$$

From (4) the normal vector is

$$\phi\rho + a^{-1} \dots\dots\dots\dots\dots\dots\dots(8);$$

therefore the equation of the normal is

$$\sigma = \rho + x (\phi\rho + a^{-1}) \dots\dots\dots\dots\dots\dots(9).$$

Equation (2) when exhibited as

$$a^2\phi\rho = \rho - a^{-1}Sa\rho,$$

reads by (6), 'vector along $NP = SP -$ vector along AN', which requires that

$$NP = a^2\phi\rho\dots\dots\dots\dots\dots\dots\dots\dots(10),$$

$$SN = a^{-1}Sa\rho ;$$

$$\text{i. e. } = aSa^{-1}\rho\dots\dots\dots\dots\dots\dots\dots(11).$$

For the subtangent AT, put xa for π in (5), and there results by (6)

$$x + Sa^{-1}\rho = 1,$$

whence $$\left(x - \frac{1}{2}\right) a = \frac{1}{2} a - aSa^{-1}\rho ;$$

i. e. vector $AT = -$ vector AN (by 11);

$$\therefore \text{ line } AT = AN ;$$

and
$$ST = xa \text{ gives}$$
$$ST^2 = (a - aSa^{-1}\rho)^2$$
$$= \frac{(a^2 - Sa\rho)^2}{a^2}$$
$$= \rho^2 \text{ by (1);}$$
$$\therefore \text{ line } ST = SP,$$

whence also the tangent bisects the angle SPQ; and SQ is perpendicular to and bisected by the tangent.

From (8)
$$y(\phi\rho + a^{-1}) = PG$$
$$= PN + NG$$
$$= -a^2\phi\rho + za \text{ (by 10);}$$
$$\therefore y = -a^2, \quad y = za^2,$$
$$z = -1,$$
$$za = -a;$$
$$\text{i.e. } NG = -SD,$$
$$\text{or line } NG = SD,$$

whence the subnormal is constant.

And vector
$$GP = -y(\phi\rho + a^{-1}) = a^2(\phi\rho + a^{-1});$$
$$\therefore \text{ vector } SQ = SD + DQ$$
$$= SD + NP$$
$$= a + a^2\phi\rho$$
$$= GP,$$

and $SQGP$ is a rhombus.

Lastly,
$$\frac{1}{2}(a + a^2\phi\rho) = \frac{1}{2}SQ$$
$$= SY$$
$$= SA + AY;$$
$$\therefore AY = \frac{1}{2}a^2\phi\rho;$$

or (10) AY is parallel to, and equal to half of NP.

53. If now we substitute Cartesian co-ordinates, making

$$\rho = xi + yj, \quad \alpha = -2ai;$$

we shall have

$$Sa^{-1}\rho = -\frac{x}{2a},$$

$$a^{-1}Sa\rho = xi,$$

$$\phi\rho = -\frac{yj}{4a^2};$$

and equation (3) becomes

$$\frac{y^2}{4a^2} - \frac{x}{a} = 1,$$

$$\text{or } y^2 = 4a\,(a + x)$$

$$= 4ax' \text{ if } x' = AN.$$

The locus of the middle points of parallel chords is thus found.

Let the chords be parallel to β, π the vector of the middle point of one of the chords,

then $\qquad\qquad\qquad \pi + x\beta = \rho,$

and $\qquad S\,(\pi + x\beta)\,\phi\,(\pi + x\beta) + 2Sa^{-1}\,(\pi + x\beta) = 1;$

which, since the term involving x must disappear, gives

$$S\pi\phi\beta + Sa^{-1}\beta = 0,$$

a straight line perpendicular to $\phi\beta$, i. e. (6) parallel to the axis.

This equation may be written

$$S\beta\,(\phi\pi + a^{-1}) = 0,$$

which shews (8) that the chords are perpendicular to the normal vector at the point where $\rho = \pi$, i. e. at the point where the locus of the chords meets the curve : in other words, the chords are parallel to the tangent at the extremity of the diameter which bisects them.

54. EXAMPLES.

Ex. 1. *If two chords be drawn always parallel to given lines, and cut one another at points either within or without the parabola, the ratio of the rectangles of their segments is always the same whatever be their point of section.*

Let POp, QOq be the chords drawn through O, and always parallel respectively to β and γ, which we will suppose to be unit vectors.

Let δ be the vector to O,

then $$\rho = \delta + x\beta$$

gives from equation (3)

$$S(\delta + x\beta)(\phi\delta + \phi x\beta + 2a^{-1}) = 1;$$
$$\therefore x^2 S\beta\phi\beta + S\delta\phi\delta + 2Sa^{-1}\delta + Ax = 1,$$

the product of the two values of x being

$$\frac{S\delta\phi\delta + 2Sa^{-1}\delta - 1}{S\beta\phi\beta};$$
$$\therefore OP.Op : OQ.Oq :: \frac{1}{S\beta\phi\beta} : \frac{1}{S\gamma\phi\gamma}$$

a constant ratio whatever be O.

Cor. Let θ, θ' be the angles in which β and γ cut the axis; then since β, γ are unit vectors, if ρ be a vector to the parabola, drawn from S parallel to POp, which we may now call SP;

$$\rho = n\beta, \quad \phi\rho = \phi(n\beta) = n\phi\beta \text{ (44. 2),}$$

will give

$$S\beta\phi\beta = \frac{S\rho\phi\rho}{n^2} = \frac{S\rho\phi\rho}{SP^2},$$

in which case $$\phi\rho \text{ is } \frac{NP}{a^2};$$

$$\therefore S\beta\phi\beta : S\gamma\phi\gamma :: \sin\theta\frac{NP}{SP} : \sin\theta'\frac{N'P'}{SP'} :: \sin^2\theta : \sin^2\theta';$$

$$\text{and, } OP.Op : OQ.Oq :: \frac{1}{\sin^2\theta} : \frac{1}{\sin^2\theta'}.$$

Ex. 2. *Find the locus of the point which divides a system of parallel chords into segments whose product is constant.*

By the last example, the equation of the locus is

$$\frac{S\delta\phi\delta + 2Sa^{-1}\delta - 1}{S\beta\phi\beta} = c,$$

a parabola similar to the given parabola.

Ex. 3. *The perpendicular from A on the tangent, and the line PQ are produced to meet in R : find the locus of R.*

By Art. 52. 8, $AR = x\,(\phi\rho + a^{-1})$,

and $PR = ya$;

$$\therefore \frac{a}{2} + x\,(\phi\rho + a^{-1}) = \rho + ya = \pi.$$

Operate by $S\phi\rho$,

and $x\,(\phi\rho)^2 = S\rho\phi\rho$

$$= a^2\,(\phi\rho)^2 \ (52.\ 7)\ ;$$

$$\therefore\ x = a^2,$$

and $\pi = \dfrac{a}{2} + a^2\,(\phi\rho + a^{-1})$

$$= \frac{3a}{2} + a^2\phi\rho \text{ is the equation required ;}$$

and, since $S\left(\pi - \dfrac{3a}{2}\right)a = 0$, it is that of a straight line perpendicular to the axis, at the distance $3a$ from S.

Ex. 4. *To find the locus of the intersection with the tangent of the perpendicular on it from the vertex.*

If π be the vector perpendicular on the tangent from A, we have by (52. 8)

$$\pi = x\,(\phi\rho + a^{-1})\dots\dots\dots\dots\dots\dots\dots(1),$$

and the equation of the tangent gives, putting $\pi + \dfrac{a}{2}$ in place of π in (52. 5), and multiplying by 2,

$$2S\pi\phi\rho + 2Sa^{-1}\pi + 2Sa^{-1}\rho = 1\dots\dots\dots\dots\dots(2),$$

we have also

$$S\rho\,(\phi\rho + 2a^{-1}) = 1\dots\dots\dots\dots\dots\dots(3).$$

From these three equations we have to eliminate x and ρ.

Equation (1) gives

$$S a \pi = x,$$

which gives x,

and

$$S \pi \phi \rho = x \, (\phi \rho)^2,$$

which substituted in (2) gives

$$2x \, (\phi \rho)^2 + 2 S a^{-1} \pi + 2 S a^{-1} \rho = 1.$$

Also, substituting (52. 7) $a^2 (\phi \rho)^2$ for $S \rho \phi \rho$, equation (3) gives

$$a^2 (\phi \rho)^2 + 2 S a^{-1} \rho = 1 \; ;$$

therefore by subtraction

$$(2x - a^2) \, (\phi \rho)^2 + 2 S a^{-1} \pi = 0,$$

i.e. $\;\; (2 S a \pi - a^2) \, (\phi \rho)^2 + 2 S a^{-1} \pi = 0,$

which from (1) becomes, multiplying by $S^2 a \pi$,

$$(2 S a \pi - a)^2 \, (\pi - a^{-1} S a \pi)^2 + 2 S^2 a \pi S a^{-1} \pi = 0.$$

This equation at once reduces to

$$2 \pi^2 S_1 \pi - \pi^2 a^2 + S^2 a \pi = 0,$$

an equation which, when $4a$ is written in place of a, becomes identical with that obtained in Art. 37, Ex. 8.

The locus is therefore a cissoid, the diameter of the generating circle being AD.

55. It will probably have suggested itself to the reader, that there exists a large class of problems to which the processes we have illustrated are scarcely if at all applicable. Hence there may have arisen a contrast between the Cartesian Geometry and Quaternions unfavourable to the latter. To remove this un-favourable impression, all that is required in a reader familiar with the older Geometry is a little experience in combining the logic of the new analysis with the forms of the old. He will then see how simple and direct are the arguments which he can bring to bear on any individual problem, and consequently how little the memory is taxed.

We propose in this Article to put the reader in the track of employing his old forms in conjunction with quaternion reasonings.

We shall work several examples on the parabola and the hyperbola. Having applied quaternions pretty fully to the ellipse in what has preceded, we will limit ourselves to a single example in this case.

1. *The Parabola.* If the unit vector along any diameter of the parabola be α, and the unit vector parallel to the tangent at its extremity be β; we may write the equation of the parabola under the form

$$\rho = x\alpha + y\beta$$

$$= \frac{y^2}{4a}\,\alpha + y\beta \quad\ldots\ldots\ldots\ldots\ldots\ldots(1).$$

For the particular case in which the diameter in question is the axis, and the tangent at its extremity parallel to the directrix

$$\rho = \frac{y^2}{4a}\,\alpha + y\beta \quad\ldots\ldots\ldots\ldots\ldots\ldots(2),$$

where a is AS (Art. 52).

This is the most convenient form when the focus is referred to.

In other cases a somewhat simpler form may be obtained by supposing α, or if necessary both α and β of equation (1) to be other than unit vectors.

The equation may then be written under the form

$$\rho = \frac{t^2}{2}\,\alpha + t\beta \quad\ldots\ldots\ldots\ldots\ldots\ldots(3).$$

To find the equation of the tangent, we have

$$\rho' = \frac{t'^2}{2}\,\alpha + t'\beta\,;$$

$$\therefore\ \rho' - \rho = (t' - t)\left(\frac{t' + t}{2}\,\alpha + \beta\right).$$

T. Q.   8

Now $\rho' - \rho$ is a vector along the secant; and its limit is a vector along the tangent: hence any vector along the tangent is a multiple of $t\alpha + \beta$; and the equation of the tangent may be written

$$\tau = \frac{t^2}{2}\,\alpha + t\beta + x\,(t\alpha + \beta) \dots\dots\dots\dots\dots(4).$$

EXAMPLES.

Ex. 1. *If AP, AQ be chords drawn at right angles to one another from A; PM, QN perpendiculars on the axis, then the latus rectum is a mean proportional between AM and AN; or between PM and QN.*

If $PM = y$, $QN = y'$,

$$AP = \frac{y^2}{4a}\,\alpha + y\beta, \quad AQ = \frac{y'^2}{4a}\,\alpha - y'\beta.$$

Now $S\,(AP \cdot AQ) = 0$ (22. 7);

$$\therefore \quad \frac{y^2 y'^2}{(4a)^2} - yy' = 0,$$

$$\text{or} \quad yy' = (4a)^2;$$

therefore also $\quad\quad\quad\quad\quad xx' = (4a)^2.$

Ex. 2. *If the rectangle of which AP, AQ are the sides be completed, the further angle will trace out a parabola similar to the given parabola, the distance between the two vertices being equal to twice the latus rectum.*

$$\rho = AP + AQ$$

$$= \frac{y^2 + y'^2}{4a}\,\alpha + (y - y')\,\beta$$

$$= \frac{(y - y')^2}{4a}\,\alpha + (y - y')\,\beta + 8a\alpha.$$

Ex. 3. *The circle described on a focal chord as diameter touches the directrix; and the circle described on any other chord does not reach the directrix.*

Let PQ be *any* chord, centre O,

$$AP = \frac{y^2}{4a}\,a + y\beta, \quad AQ = \frac{y'^2}{4a}\,a + y'\beta.$$

The equation of the circle with centre O, radius OP, is

$$\left(\rho - \frac{AQ + AP}{2}\right)^2 = \left(\frac{AQ - AP}{2}\right)^2,$$

or $\quad \rho^2 - S(AP + AQ)\rho + S(AP \cdot AQ) = 0.$

At the points in which this circle meets the directrix

$$\rho = -a\alpha + z\beta\,;$$

$$\therefore -a^2 - z^2 - \frac{y^2 + y'^2}{4} + z(y + y') - \frac{y^2 y'^2}{(4a)^2} - yy' = 0,$$

or $\qquad z^2 - z(y + y') + \frac{(y + y')^2}{4} = -\left(\frac{yy'}{4a} + a\right)^2.$

This equation is possible only when

$$yy' + 4a^2 = 0\,;$$

i.e. when the chord is a focal chord.

In this case the two values of z are equal, each being $\dfrac{y + y'}{2}$; and the directrix is a tangent to the circle.

Ex. 4. *Two parabolas have a common focus and axis; their vertices are turned in opposite directions. A focal chord cuts them in PQ, $P'Q'$, so that $PP'SQQ'$ are in order. Prove* (1) *that $SP \cdot SP' = SQ \cdot SQ'$;* (2) *that $SP : SQ$ is a constant ratio; and* (3) *that the tangents at P, P' are at right angles to one another.*

The equations of the parabolas are

$$\rho = -a\alpha + \frac{y^2}{4a}\,a + y\beta,$$

$$\rho' = a'\alpha - \frac{y'^2}{4a'}\,a + y'\beta,$$

the focus being the origin.

Now since ρ, ρ' are in the same straight line when the common chord is the focal chord, we have

$$\rho' = p\rho;$$

$$\therefore\ a' - \frac{y'^2}{4a'} = -pa + \frac{py^2}{4a},$$

$$y' = py,$$

$$\therefore\ (yy' - 4aa')(a'y + ay') = 0.$$

Taking the former factor, we must have y, y' on the same side of the axis with a constant product; therefore

$$SP \cdot SP' = SQ \cdot SQ'.$$

The second factor gives $SP : SQ'$ a constant ratio $a : a'$.

Lastly, by Equation (4), the tangent vectors at P and P' are parallel to

$$\frac{y}{2a}\,a + \beta,\quad -\frac{y'}{2a'}\,a + \beta.$$

Now $\quad S\left(\frac{y}{2a}\,a + \beta\right)\left(-\frac{y'}{2a'}\,a + \beta\right) = \frac{yy'}{4aa'} - 1 = 0;$

therefore the tangents are at right angles to one another.

Ex. 5. *If a triangle be inscribed in a parabola, the three points in which the sides are met by the tangents at the angles lie in a straight line.*

Let OPQ be the triangle.

Take O as the origin, then

$$\rho = \frac{t^2}{2}\,a + t\beta,$$

$$\rho' = \frac{t'^2}{2}\,a + t'\beta,$$

$$\pi = \frac{t^2}{2}\,a + t\beta + x(ta + \beta),$$

$$\pi' = \frac{t'^2}{2}\,a + t'\beta + x'(t'a + \beta),$$

are the vectors OP, OQ, and the equations of the tangents at P and Q.

If QO meet in A the tangent at P,

$$OA = \frac{t^2}{2}\,\alpha + t\beta + x\,(t\alpha + \beta)$$

$$= y\,OQ$$

$$= y\left(\frac{t'^2}{2}\,\alpha + t'\beta\right);$$

$$\therefore \frac{t^2}{2} + tx = \frac{t'^2}{2}\,y,$$

$$t + x = t'y,$$

$$y = \frac{t^2}{2tt' - t'^2},$$

and

$$OA = \frac{t^2}{2tt' - t'^2}\left(\frac{t'^2}{2}\,\alpha + t'\beta\right)$$

$$= \frac{t^2}{2t - t'}\left(\frac{t'}{2}\,\alpha + \beta\right).$$

Similarly if the tangent at Q meets PO in B,

$$OB = \frac{t'^2}{2t' - t}\left(\frac{t}{2}\,\alpha + \beta\right).$$

If the tangent at O meets PQ in C,

$$OC = OP + z\,(PQ)$$

$$= OP + z\,(OQ - OP)$$

$$= \frac{t^2}{2}\,\alpha + t\beta + z\left\{\frac{t'^2 - t^2}{2}\,\alpha + (t' - t)\beta\right\}.$$

But

$$OC = v\beta;$$

$$\therefore \frac{t^2}{2} + z\,\frac{t'^2 - t^2}{2} = 0,$$

$$t + z\,(t' - t) = v,$$

$$v = \frac{tt'}{t + t'},$$

and $$OC = \frac{tt'}{t + t'}\,\beta.$$

Now $$\frac{2t - t'}{t}\,OA - \frac{2t' - t}{t'}\,OB - \frac{t^2 - t'^2}{tt'}\,OC = 0,$$

and also $$\frac{2t - t'}{t} - \frac{2t' - t}{t'} - \frac{t^2 - t'^2}{tt'} = 0\,;$$

therefore (Art. 13) A, B, C are in a straight line.

2. *The ellipse.* If a, β are unit vectors along the axes, the equation of the ellipse may be written

$$\rho = xa + y\beta,$$

where $$y^2 = \frac{b^2}{a^2}\,(a^2 - x^2) = m\,(a^2 - x^2)\,;$$

and the equation of the tangent will be readily seen to be

$$\pi = xa + y\beta + X\,(ya - mx\beta).$$

A single example will suffice.

Ex. *If tangents be drawn at three points P, Q, R of an ellipse intersecting in R', Q', P', prove that*

$$PR'.\,QP'.\,RQ' = PQ'.\,QR'.\,RP'.$$

If x, y; x', y'; x'', y'' are respectively the co-ordinates of P, Q, R; we shall have

$$CR' = xa + y\beta + X\,(ya - mx\beta)$$

$$= x'a + y'\beta + X'\,(y'a - mx'\beta)\,;$$

$$\therefore\ x + Xy = x' + X'y',$$

$$y - mXx = y' - mX'x'\,;$$

$$\therefore\ mX\,(x'y - y'x) = mx'^2 + y'^2 - mxx' - yy'$$

$$= b^2 - mxx' - yy'.$$

Hence $$mX'\,(xy' - x'y) = b^2 - mxx' - yy'$$

$$= -mX\,(xy' - x'y)\,;$$

$$\therefore\ X = -X',$$

$$Y = -Y'\ \text{for}\ Q',$$

$$Z = -Z'\ \text{for}\ P',$$

and $$XY'Z = -X'YZ'.$$

Now
$$\frac{X}{Y} = \frac{PR'}{PQ'}, \quad \&c.$$

hence the proposition.

3. *The hyperbola.* If α, β are unit vectors parallel to the asymptotes CX, CY, the equation of the hyperbola may be written

$$\rho = x\alpha + y\beta$$
$$= x\alpha + \frac{C}{x}\beta,$$

since
$$xy = \frac{a^2 + b^2}{4} = C.$$

If α, β be not both units we may write the equation under the simpler form

$$\rho = t\alpha + \frac{\beta}{t} \quad \dots\dots\dots\dots\dots\dots\dots(1).$$

To find the equation of the tangent, we have as usual a vector parallel to the secant

$$= \rho' - \rho = (t' - t)\left(\alpha - \frac{\beta}{tt'}\right)$$
$$= \left(\frac{t' - t}{t}\right)\left(t\alpha - \frac{\beta}{t'}\right),$$

and a vector parallel to the tangent will be

$$t\alpha - \frac{\beta}{t} \quad \dots\dots\dots\dots\dots\dots\dots\dots(2).$$

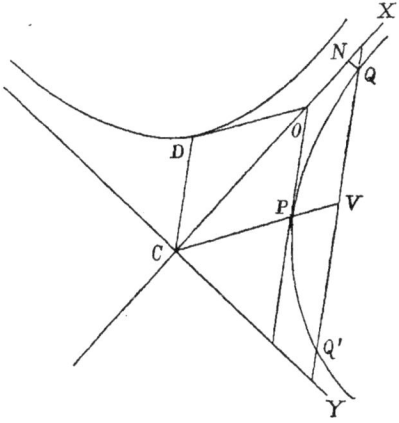

Hence the equation of the tangent is

$$\pi = ta + \frac{\beta}{t} + x \left(ta - \frac{\beta}{t} \right) \dots\dots\dots\dots\dots (3).$$

Cor. It is evident that

$$ta + \frac{\beta}{t}, \quad ta - \frac{\beta}{t};$$

are conjugate semi-diameters.

Examples.

Ex. 1. *One diagonal of a parallelogram whose sides are the co-ordinates being the radius vector, the other diagonal is parallel to the tangent.*

We have $\qquad CN = ta, \quad \ddot{N}Q = \frac{\beta}{t},$

$$CQ = ta + \frac{\beta}{t},$$

and the other diagonal is

$$ta - \frac{\beta}{t},$$

which, equation (2), is parallel to the tangent at Q.

Ex. 2. *Any diameter CP bisects all the chords which are parallel to the tangent at P.*

Let CP be $\qquad ta + \frac{\beta}{t},$

then the tangent at P is parallel to

$$ta - \frac{\beta}{t};$$

$$\therefore \quad CQ = CV + VQ = X \left(ta + \frac{\beta}{t} \right) + Y \left(ta - \frac{\beta}{t} \right).$$

But as Q is a point in the hyperbola, this equation must have the form

$$CQ = Ta + \frac{\beta}{T};$$

$$\therefore (X + Y)\,t = T,$$

$$(X - Y)\frac{1}{t} = \frac{1}{T},$$

and
$$X^2 - Y^2 = 1,$$

an equation which gives two equal values of Y with opposite signs, for every value of X.

Hence all chords are bisected.

Cor. $X^2 - Y^2 = 1$ is

$$\left(\frac{CV}{CP}\right)^2 - \left(\frac{QV}{CD}\right)^2 = 1,$$

CD being
$$ta - \frac{\beta}{t} = PO.$$

This is the ordinary equation of the hyperbola referred to conjugate diameters.

Ex. 3.   *If TQ, $T'Q'$ be two tangents to the hyperbola intersecting in R and terminated at T, T', Q, Q' by the asymptotes; then (1) TQ' is parallel to $T'Q$; (2) area of triangle TRT' = area of triangle QRQ', and (3) CR bisects TQ' and $T'Q$.*

The equation of the tangent

$$\pi = ta + \frac{\beta}{t} + x\left(ta - \frac{\beta}{t}\right),$$

gives
$$CT = 2ta,$$

(the coefficient of β being 0),

$$CQ = \frac{2\beta}{t},$$

$$CT' = 2t'a,$$

$$CQ' = \frac{2\beta}{t'};$$

$$\therefore\ Q'T = 2at - \frac{2\beta}{t'} = \frac{2}{t'}(att' - \beta),$$

$$QT' = \frac{2}{t}(att' - \beta);$$

therefore $Q'T$ is parallel to QT'.

Again, $CR = CQ + QR = CQ + x\,(CT - CQ)$

$$= \frac{2\beta}{t} + x2\left(at - \frac{\beta}{t}\right).$$

Also $CR = \frac{2\beta}{t'} + x'2\left(at' - \frac{\beta}{t'}\right);$

$$\therefore\ xt = x't',$$

$$\frac{1}{t} - \frac{x}{t} = \frac{1}{t'} - \frac{x'}{t'},$$

$$x = \frac{t'}{t + t'},$$

$$x' = \frac{t}{t + t'},$$

and $xx' = (1 - x)\,(1 - x'),$

i.e. $QR\,.\,Q'R = RT\,.\,RT',$

and the triangles TRT', QRQ' are equal.

Lastly, $CR = \frac{2\beta}{t} + \frac{2t'}{t + t'}\left(at - \frac{\beta}{t}\right)$

$$= \frac{t'}{t + t'}\left(2ta + \frac{2\beta}{t'}\right),$$

or CR is in the direction of the diagonal of the parallelogram of which the sides are CT, CQ'; and therefore CR bisects TQ' and $T'Q$.

Ex. 4. *If through Q, P, Q' parallels be drawn to CX meeting CY in E, F, G; CE, CF, CG are in continued proportion.*

$$CP = ta + \frac{\beta}{t};$$

$$Q'Q = m\left(ta - \frac{\beta}{t}\right);$$

$$\therefore\ CQ = CV + VQ$$

$$= X\left(ta + \frac{\beta}{t}\right) + Y\left(ta - \frac{\beta}{t}\right),$$

$$CQ' = X\left(t\alpha + \frac{\beta}{t}\right) - Y\left(t\alpha - \frac{\beta}{t}\right),$$

$$CE = (X - Y)\frac{\beta}{t},$$

$$CF = \frac{\beta}{t},$$

$$CG = (X + Y)\frac{\beta}{t};$$

and $\qquad CE \cdot CG = CF^2;$

because $\qquad X^2 - Y^2 = 1$ (Ex. 2).

Ex. 5. *If a chord of a hyperbola be one diagonal of a parallelogram whose sides are parallel to the asymptotes, the other diagonal passes through the centre.*

Let the chord be PQ; ρ, ρ' the vectors to P and Q; then

$$QP = \rho - \rho' = \alpha t + \frac{\beta}{t} - \left(\alpha t' + \frac{\beta}{t'}\right).$$

Now when one diagonal of a parallelogram is $m\alpha + n\beta$, the other will be $m\alpha - n\beta$.

Therefore in the case before us, the other diagonal is

$$\alpha(t - t') - \beta\left(\frac{1}{t} - \frac{1}{t'}\right)$$

$$= (t - t')\left(\alpha + \frac{\beta}{tt'}\right)$$

$$= \frac{t - t'}{t + t'}\left\{\alpha(t + t') + \beta\left(\frac{1}{t} + \frac{1}{t'}\right)\right\}$$

$$= \frac{t - t'}{t + t'}(\rho + \rho').$$

And it is therefore in the same straight line with the line which joins the centre of the hyperbola with the middle point of PQ; whence the truth of the proposition.

Ex. 6. *If two tangents to a hyperbola at the extremities Q, Q′ of a diameter, meet a tangent at P in the points T, T′; and if CD, CD′ are the semi-diameters conjugate to CP, CQ; then* (1) $PT : QT :: PT' : Q'T' :: CD : CD';$

and (2) $PT . PT' = CD^2.$

If $t, t', -t'$, correspond to P, Q, Q', then

$$CT = at + \frac{\beta}{t} + x\left(at - \frac{\beta}{t}\right)$$

$$= at' + \frac{\beta}{t'} + x'\left(at' - \frac{\beta}{t'}\right),$$

gives $t + xt = t' + x't',$

$$\frac{1}{t} - \frac{x}{t} = \frac{1}{t'} - \frac{x'}{t'},$$

$$x = \frac{t' - t}{t' + t} = -x'.$$

Similarly $$CT' = at + \frac{\beta}{t} + y\left(at - \frac{\beta}{t}\right)$$

$$= -at' - \frac{\beta}{t'} - y'\left(at' - \frac{\beta}{t'}\right),$$

gives $t + yt = -t' - y't',$

$$\frac{1}{t} - \frac{y}{t} = -\frac{1}{t'} + \frac{y'}{t'},$$

whence $y = \frac{t' + t}{t' - t} = -y'.$

Now $x : y :: x' : y'$

gives $PT : QT :: PT' : QT'$

$$:: CD : CD'.$$

And $xy = 1$

gives $PT . PT' = CD^2.$

Cor. $x'y' = 1,$

gives $QT . Q'T' = CD'^2.$

Ex. 7.   *Straight lines move so that the triangular area which they cut off from two given straight lines which meet one another is constant : to find the locus of their ultimate intersections.*

Let OAA', OBB' be the fixed lines, AB, $A'B'$ two of the moving lines with the condition that

$$OA . OB = OA' . OB'.$$

If a, β be unit vectors along OA, OB,

$$OA = ta, \quad OB = u\beta; \quad OA' = t'a, \quad OB' = u'\beta,$$

the point of intersection of AB, AB' gives

$$\rho = ta + x\,(u\beta - ta)$$

$$= t'a + x'\,(u'\beta - t'a),$$

$$\therefore \quad xu = x'u',$$

and
$$t\,(1-x) = t'\,(1-x')$$

$$= t'\left(1 - \frac{xu}{u'}\right).$$

Now $tu = t'u' = c$ because the triangle has a constant area;

$$\therefore \quad x = \frac{t}{t+t'} = \frac{1}{2} \text{ ultimately;}$$

$$\therefore \quad \rho = \frac{1}{2}ta + \frac{1}{2}u\beta = \frac{1}{2}ta + \frac{1}{2}\frac{c\beta}{t},$$

the equation of a hyperbola.

ADDITIONAL EXAMPLES TO CHAP. VII.

1.   In the parabola $SY^2 = SP . SA$.

2.   If the tangent to a parabola cut the directrix in R, SR is perpendicular to SP.

3.   A circle has its centre at the vertex A of a parabola whose focus is S, and the diameter of the circle is $3AS$.   Prove that the common chord bisects AS.

4.   The tangent at any point of a parabola meets the directrix and latus rectum in two points equally distant from the focus.

5. The circle described on SP as diameter is touched by the tangent at the vertex.

6. Parabolas have their axes parallel and all pass through two given points. Prove that their foci lie in a conic section.

7. Two parabolas have a common directrix. Prove that their common chord bisects at right angles the line joining their foci.

8. The portion of any tangent to the parabola between tangents which meet in the directrix subtends a right angle at the focus.

9. If from the point of contact of a tangent to a parabola a chord be drawn, and another line be drawn parallel to the axis meeting the chord, tangent and curve ; this line will be divided by them in the same ratio as it divides the chord.

10. The middle points of focal chords describe a parabola whose latus rectum is half that of the given parabola.

11. PSQ is a focal chord of a parabola: PA, QA meet the directrix in y, z. Prove that Pz, Qy are parallel to the axis.

12. The tangent at D to the conjugate hyperbola is parallel to CP.

13. The portion of the tangent to a hyperbola which is intercepted by the asymptotes is bisected at the point of contact.

14. The locus of a point which divides in a given ratio lines which cut off equal areas from the space enclosed by two given straight lines is a hyperbola of which these lines are the asymptotes.

15. The tangent to a hyperbola at P meets an asymptote in T, and TQ is drawn to the curve parallel to the other asymptote. PQ produced both ways meets the asymptotes in R, R' : RR' is trisected in P, Q.

16. From any point R of an asymptote, RN, RM are drawn parallel to conjugate diameters intersecting the hyperbola and its conjugate in P and D. Prove that CP and CD are conjugate.

17. The intercepts on any straight line between the hyperbola and its asymptotes are equal.

18. If QQ' meet the asymptotes in R, r,

$$RQ \cdot Qr = PO^2.$$

19. If the tangent at any point meet the asymptotes in X and Y, the area of the triangle XCY is constant.

CHAPTER VIII.

CENTRAL SURFACES OF THE SECOND ORDER, PARTICULARLY THE ELLIPSOID AND CONE.

56. *The Ellipsoid.* In discussing central surfaces of the second order, we shall speak as if our results were limited to the ellipsoid. That such limitation is not, in most cases, necessarily imposed on us, will be apparent to any one who has a slender acquaintance with ordinary Analytical Geometry. We adopt it in order that our language may have more precision, and that, in some instances, our analysis may have greater simplicity. If the centre be made the origin it is clear that the scalar equation can contain no such term as $ASa\rho$, for the *definition* of a central surface requires that the equation shall be satisfied both by $+\rho$ and by $-\rho$.

If we turn to the equation of the ellipse (Art. 43), we shall see at once that the equation of the ellipsoid must have the form

$$a\rho^2 + bS^2a\rho + 2cSa\rho S\beta\rho + \ldots = 1.$$

Now if, as in the Article referred to, we put

$$\phi\rho = a\rho + baSa\rho + c\,(aS\beta\rho + \beta Sa\rho) + \ldots$$

we shall have

$$S\rho\phi\rho = a\rho^2 + bS^2a\rho + 2cSa\rho S\beta\rho + \ldots$$
$$= 1,$$

the equation required.

It will be seen that, as in Arts. 32, 33, one form of the equation of the straight line was found to coincide exactly with the equation of a plane, so a form of the equation of the ellipse coincides exactly with the equation of the ellipsoid.

It is evident that the three properties of $\phi\rho$ given in Art. 44 are true of $\phi\rho$ in its present form.

57. To find the equation of the tangent plane.

Let a secant plane pass through the point whose vector is ρ; and let ρ' be the vector to any point of section.

Put $\rho' = \rho + \beta$, where β is a vector along the secant plane;

then $\qquad S\rho'\phi\rho' = S(\rho + \beta)\phi(\rho + \beta).$

Hence, observing that (44)

$$\phi(\rho + \beta) = \phi\rho + \phi\beta,$$

and $\qquad\qquad S\rho\phi\beta = S\beta\phi\rho,$

we have $\qquad S\rho'\phi\rho' = S\rho\phi\rho + 2S\beta\phi\rho + S\beta\phi\beta;$

$\qquad\qquad$ i.e. $2S\beta\phi\rho + S\beta\phi\beta = 0.$

Now (45), as the secant plane approaches the tangent plane, the sum of these two expressions approaches in value to the first alone: that is, for the tangent plane, $S\beta\phi\rho = 0$, where β is a vector along that plane.

If π be the vector to a point in the tangent plane,

$$\pi = \rho + x\beta;$$
$$\therefore \ S(\pi - \rho)\phi\rho = xS\beta\phi\rho$$
$$= 0,$$
and $\qquad\qquad S\pi\phi\rho = S\rho\phi\rho$
$$= 1$$

is the equation of the tangent plane.

COR. $\phi\rho$ is a vector perpendicular to the tangent plane at the extremity of the vector ρ.

58. If OY be perpendicular from the centre O on the tangent plane; then, since $\phi\rho$ is a vector perpendicular to that plane, $OY = x\phi\rho$ and $Sx(\phi\rho)^2 = 1$, giving

$$OY = T(x\phi\rho) = T\frac{1}{\phi\rho}.$$

Sir W. Hamilton terms $\phi\rho$ the *vector of proximity*. [In fact vector $OY = (\phi\rho)^{-1}$.]

T. Q. $\qquad\qquad\qquad\qquad\qquad\qquad\qquad\qquad$ 9

59. If tangent planes all pass through a fixed point, the curve of contact is a plane curve.

Let T be the fixed point; vector a; ρ the vector to a point of contact.

Then (Art. 57) $\qquad S a \phi \rho = 1$;

$$\text{i. e. } S \rho \phi a = 1 \ (44. \ 3),$$

which is the equation in ρ of a plane perpendicular to ϕa.

Now ϕa is the normal vector of the point where OT cuts the ellipsoid;

$\therefore$ the curve of contact lies in a plane parallel to the tangent plane at the extremity of the diameter drawn to the given point.

The plane of contact is called the polar plane to the point.

60. Tangent planes are all parallel to a given straight line, to find the curve of contact.

Let a be a vector parallel to the given line; then

$$\pi = \rho + x a$$

is a point in the tangent plane;

$$\therefore S (\rho + x a) \phi \rho = 1 \ ;$$

and $\qquad\qquad S a \phi \rho = 0,$

or $\qquad\qquad S \rho \phi a = 0,$

the equation of a plane through the origin perpendicular to ϕa: that is, the curve of contact lies in a plane through the centre parallel to the tangent plane at the extremity of the diameter which is parallel to the given line.

61. To find the locus of the middle points of parallel chords.

Let each of the chords be parallel to a, π the vector to the middle point of one of them; then $\pi + x a$, $\pi - x a$ are points in the ellipsoid.

From the first,

$$S (\pi + x a) \phi (\pi + x a) = 1 \ (\text{Art. 56}) ;$$

$$\text{i. e. } S \pi \phi \pi + 2 x S \pi \phi a + x^2 S a \phi a = 1.$$

From the second,

$$S\pi\phi\pi - 2xS\pi\phi a + x^2 Sa\phi a = 1;$$

$$\therefore \text{ subtracting, } S\pi\phi a = 0 \quad (1),$$

i.e. the locus is a plane through the centre perpendicular to ϕa, or parallel to the tangent plane at the extremity A of the diameter which is drawn parallel to a.

If we call this the plane BOC, B and C being any points in which it cuts the ellipsoid; and if $OB = \beta$, $OC = \gamma$, we shall have

$$S\beta\phi a = 0, \quad S\gamma\phi a = 0,$$

and therefore $\qquad\qquad Sa\phi\beta = 0,$

or a satisfies the equation $\quad S\pi\phi\beta = 0$

of the plane which bisects all chords parallel to OB (Equation 1).

Let AOC be this plane which bisects all chords parallel to OB.

Then, since OC or γ is a vector in it,

$$S\gamma\phi\beta = 0, \text{ i.e. } S\beta\phi\gamma = 0.$$

But we have already proved that

$$S\gamma\phi a = 0, \text{ i.e. } Sa\phi\gamma = 0,$$

because γ is in the plane BOC;

$\therefore$ by equation (1) a, β both satisfy the equation of the plane $S\pi\phi\gamma = 0$, which is the plane bisecting all chords parallel to γ; that plane is therefore the plane AOB: we are thus presented with three lines OA, OB, OC such that all chords parallel to any one of them are bisected by the diametral plane which passes through the other two.

We may term these lines *conjugate semi-diameters*, and the corresponding diametral planes *conjugate diametral planes*.

It is evident that the number of conjugate diameters is unlimited.

COR. We have the following equations:

$$Sa\phi\beta = 0 = S\beta\phi a,$$

$$S\beta\phi\gamma = 0 = S\gamma\phi\beta,$$

$$Sa\phi\gamma = 0 = S\gamma\phi a \quad (2).$$

9—2

They shew that γ is perpendicular to both ϕa and $\phi \beta$, and is therefore a vector perpendicular to their plane; hence, as in 34. 4,

$$\gamma = x V \phi a \phi \beta.$$

In the same way, since $\phi \gamma$ is perpendicular to both a and β, we have

$$\phi \gamma = y V a \beta ;$$

or, neglecting tensors, we have the following vector equalities :

$$\gamma = V \phi a \phi \beta, \quad \beta = V \phi a \phi \gamma, \quad a = V \phi \beta \phi \gamma,$$
$$\phi \gamma = V a \beta, \quad \phi \beta = V a \gamma, \quad \phi a = V \beta \gamma \quad (3).$$

Note also

$$y \phi^{-1} V a \beta = x V \phi a \phi \beta,$$

upon which Hamilton founded his solution of linear equations.

62. If as in Art. 47 we write $- \psi \psi \rho$ for $\phi \rho$, $\psi \rho$ being still a vector, the equation of the ellipsoid assumes the form

$$S \rho \psi (\psi \rho) = -1,$$
$$\text{i.e. (44)} \quad S \psi \rho \psi \rho = -1$$
$$(\psi \rho)^2 = - T (\psi \rho)^2 = -1 \ldots \ldots \ldots (1),$$

which, if we put $\sigma = \psi \rho$, becomes $T \sigma = 1$, the equation of a sphere.

Hence the ellipsoid can be changed into the sphere and *vice versâ*, by a linear deformation of each vector, the operator being the function ψ or its inverse.

The equations

$$S a \phi \beta = 0, \ \&c.$$

now become

$$S a \psi^2 \beta = 0,$$
$$\text{i.e.} \ S \psi a \psi \beta = 0, \ \&c., \ \&c. \ldots \ldots \ldots \ldots (2).$$

(1) and (2) shew that ψa, $\psi \beta$, $\psi \gamma$ are unit vectors at right angles to one another.

If we term the sphere $T \sigma = 1$ the unit-sphere, we may enunciate this result by saying that the vectors of the unit-sphere which correspond to semi-conjugate diameters form a rectangular system.

63. Let us now take i, j, k unit vectors along the principal axes of x, y, z; then we shall have

$$\rho = xi + yj + zk \dots\dots\dots\dots\dots\dots(1),$$

$$\therefore Si\rho = -x, \ \&c.$$

so that for the sake of transformations in which it is desirable that the form of ρ should be retained, we may write

$$\rho = -(iSi\rho + jSj\rho + kSk\rho)\dots\dots\dots\dots(2);$$

and as $\phi\rho$ is a linear and vector function of ρ, its vector portions along the principal axes will be multiples of

$$iSi\rho, \ jSj\rho, \ kSk\rho;$$

we may therefore write

$$\phi\rho = \frac{iSi\rho}{a^2} + \frac{jSj\rho}{b^2} + \frac{kSk\rho}{c^2}\dots\dots\dots\dots(3),$$

the form a^2 having been assumed in order to make the equation

$$S\rho\phi\rho = 1$$

coincide with the Cartesian equation

$$\frac{x^2}{a^2} + \frac{y^2}{b^2} + \frac{z^2}{c^2} = 1.$$

As
$$\phi\rho = -\psi\psi\rho \dots\dots\dots\dots\dots\dots(4),$$

we require to take $\psi\rho$ so that performing the operation ψ twice on ρ shall give the same result (with a $-$ sign) as performing the operation ϕ once.

Now a comparison of equations (2) and (3) will shew that the latter operation introduces $\dfrac{1}{a^2}$ &c. into ρ; it is evident therefore that the former operation (ψ) is to introduce $\dfrac{1}{a}$ &c. or

$$\psi\rho = -\left(\frac{iSi\rho}{a} + \frac{jSj\rho}{b} + \frac{kSk\rho}{c}\right)\dots\dots\dots\dots(5).$$

It may perhaps be worth while to verify this result. We have

$$\psi\psi\rho = -\left(\frac{iSi\psi\rho}{a} + \frac{jSj\psi\rho}{b} + \frac{kSk\psi\rho}{c}\right)$$

$$= iS\frac{i}{a}\left(\frac{iSi\rho}{a} + \frac{jSj\rho}{b} + \frac{kSk\rho}{c}\right) + \dots$$

$$= i\,\frac{i^2Si\rho}{a^2} + \dots$$

$$= -\left(\frac{iSi\rho}{a^2} + \frac{jSj\rho}{b^2} + \frac{kSk\rho}{c^2}\right)$$

$$= -\phi\rho.$$

$$\phi^2\rho = \phi\phi\rho = \frac{iSi\phi\rho}{a^2} + \frac{jSj\phi\rho}{b^2} + \frac{kSk\phi\rho}{c^2}$$

$$= -\left(\frac{iSi\rho}{a^4} + \frac{jSj\rho}{b^4} + \frac{kSk\rho}{c^4}\right)\dots\dots\dots\dots\dots(6),$$

$$\phi^{-1}\rho = a^2iSi\rho + b^2jSj\rho + c^2kSk\rho\dots\dots\dots\dots\dots(7),$$

because $\phi\phi^{-1}\rho$ produces ρ.

$$\psi^{-1}\rho = -(aiSi\rho + bjSj\rho + ckSk\rho)\dots\dots\dots\dots(8),$$

$$\rho = \psi^{-1}\psi\rho = -(aiSi\psi\rho + bjSj\psi\rho + ckSk\psi\rho)\dots\dots\dots(9).$$

It is evident that the properties of Art. 44 apply to all these functions.

64.   EXAMPLES.

Ex. 1. *Find the point on an ellipsoid, the tangent plane at which cuts off equal portions from the axes.*

Let x, y, z be the co-ordinates of the point, p the portion cut off, then

$$\rho = xi + yj + zk.$$

Now pi, pj, pk are points on the tangent plane;

$$\therefore\ Spi\phi\rho = 1,$$

which gives

$$pSi\left(\frac{iSi\rho}{a^2} + \dots\right) = 1,$$

or
$$\frac{px}{a^2} = 1.$$

Similarly
$$\frac{py}{b^2} = 1,$$

$$\frac{pz}{c^2} = 1,$$

$$\frac{x}{a^2} = \frac{y}{b^2} = \frac{z}{c^2} = \frac{1}{p} = \frac{1}{\sqrt{a^2 + b^2 + c^2}}.$$

Ex. 2. *To find the perpendicular from the centre of the ellipsoid on a tangent plane.*

$$OY^2 = \left(T\frac{1}{\phi\rho} \right)^2; \quad \text{(Art. 58)}$$

$$\therefore \frac{1}{OY^2} = (T\phi\rho)^2 = -(\phi\rho)^2 = \frac{x^2}{a^4} + \frac{y^2}{b^4} + \frac{z^2}{c^4} \quad \text{(Art. 63, 1. 3).}$$

Ex. 3. *To find the locus of the points of contact of tangent planes which make a given angle with the axis of z.*

We have
$$SkU(\phi\rho) = p,$$

$$Sk\phi\rho = pT\phi\rho,$$

or
$$\frac{z^2}{c^4} = p^2\left(\frac{x^2}{a^4} + \frac{y^2}{b^4} + \frac{z^2}{c^4} \right),$$

the equation of a cone whose axis is that of z and guiding curve an ellipse whose semi-axes are a^2, b^2.

The intersection of this surface with the ellipsoid is the locus required.

Ex. 4. *To find the locus of a point when the perpendicular from the centre on its polar plane is of constant length.*

Let π be the vector to the point, then

$S\rho\phi\pi = 1$ is the equation of the polar plane (Art. 59),

and $T\dfrac{1}{\phi\pi}$ is the length of the perpendicular on it (Art. 58);

$$\therefore S(\phi\pi)^2 = -C^2, \text{ by the question.}$$

But since (44)

$$S\delta\phi\pi = S\pi\phi\delta,$$

if δ be $\phi\pi$,

$$S\phi\pi\phi\pi = S\pi\phi\phi\pi = S\pi\phi^2\pi \ ;$$

$$\therefore S\pi\phi^2\pi = -C^2 \text{ is the equation required} \ ;$$

hence the Cartesian equation is (63. 6)

$$\frac{x^2}{a^4} + \frac{y^2}{b^4} + \frac{z^2}{c^4} = C^2.$$

Ex. 5. *The sum of the squares of three conjugate semi-diameters is constant.*

Let a, β, γ be the semi-diameters; ψa, $\psi\beta$, $\psi\gamma$ are rectangular unit vectors (Art. 62).

Now $a = -(aiSi\psi a + bjSj\psi a + ckSk\psi a)$ (63. 9) ;

$$\therefore (Ta)^2 = -a^2 = a^2 (Si\psi a)^2 + b^2 (Sj\psi a)^2 + c^2 (Sk\psi a)^2,$$

$$(T\beta)^2 = \qquad a^2 (Si\psi\beta)^2 + b^2 (Sj\psi\beta)^2 + c^2 (Sk\psi\beta)^2,$$

$$(T\gamma)^2 = \qquad a^2 (Si\psi\gamma)^2 + b^2 (Sj\psi\gamma)^2 + c^2 (Sk\psi\gamma)^2 :$$

adding, and observing that

$$(Si\psi a)^2 + (Si\psi\beta)^2 + (Si\psi\gamma)^2 = 1 \quad (31. \text{ Cor.}),$$

we get

$$(Ta)^2 + (T\beta)^2 + (T\gamma)^2 = a^2 + b^2 + c^2,$$

$$\text{i.e. } a'^2 + b'^2 + c'^2 = a^2 + b^2 + c^2.$$

Ex. 6. *The sum of the squares of the three perpendiculars from the centre on three tangent planes at right angles to one another is constant.*

We have

$$\rho = \phi^{-1}\phi\rho = a^2iSi\phi\rho + b^2jSj\phi\rho + c^2kSk\phi\rho \ (63. 7),$$

and $\phi\rho = -(iSi\phi\rho + jSj\phi\rho + kSk\phi\rho)$ (63. 2) ;

$$\therefore S\rho\phi\rho = 1 = a^2 (Si\phi\rho)^2 + b^2 (Sj\phi\rho)^2 + c^2 (Sk\phi\rho)^2$$

$$= (T\phi\rho)^2 \{a^2 (SiU\phi\rho)^2 + b^2 (Sj U\phi\rho)^2 + c^2 (SkU\phi\rho)^2\} \ ;$$

hence if ρ, ρ', ρ'' be three vectors so that $\phi\rho$, $\phi\rho'$, $\phi\rho''$ are at right angles to each other; that is, so that the tangent planes at their extremities are at right angles to one another (57. Cor.),

$$\frac{1}{(T\phi\rho)^2} + \frac{1}{(T\phi\rho')^2} + \frac{1}{(T\phi\rho'')^2}$$
$$= a^2 \{(SiU\phi\rho)^2 + (SiU\phi\rho')^2 + (SiU\phi\rho'')^2\}$$
$$+ b^2 \{(SjU\phi\rho)^2 + \dots\} + \dots$$
$$= a^2 + b^2 + c^2 \quad (31. \text{ Cor.}).$$

But $\dfrac{1}{(T\phi\rho)^2}$, &c. are the perpendiculars from the centre on the tangent planes at ρ, ρ', ρ'' (58). Hence the proposition.

Ex. 7. *The sum of the squares of the projections of three conjugate diameters on any of the principal axes is equal to the square of that axis.*

Let a, β, γ be conjugate semi-diameters; then, since

$$a = - (aiSi\psi a + bjSj\psi a + ckSk\psi a) \quad (63. \ 9),$$
$$Sia = aSi\psi a.$$

Similarly, $\qquad Si\beta = aSi\psi\beta,$
$$Si\gamma = aSi\psi\gamma \ ;$$
$$\therefore \ (Sia)^2 + (Si\beta)^2 + (Si\gamma)^2 = a^2 \{(Si\psi a)^2 + (Si\psi\beta)^2 + (Si\psi\gamma)^2\}$$
$$= a^2 (31. \text{ Cor.}),$$

because ψa, $\psi\beta$, $\psi\gamma$ are at right angles to one another (62).

But $- Sia$ is the projection of Ta along the axis of x; and similarly of the others. Hence the proposition.

Ex. 8. *The sum of the reciprocals of the squares of the three perpendiculars from the centre on tangent planes at the extremities of conjugate diameters is constant.*

Let Oy_1, Oy_2, Oy_3 be the perpendiculars.

$$\frac{1}{Oy_1^2} = - (\phi a)^2 \quad (58)$$
$$= \frac{(Sia)^2}{a^4} + \frac{(Sja)^2}{b^4} + \frac{(Ska)^2}{c^4} \quad (63. \ 3) \ ;$$

$$\frac{1}{Oy_2^{\,2}} = \frac{(Si\beta)^2}{a^4} + \frac{(Sj\beta)^2}{b^4} + \frac{(Sk\beta)^2}{c^4} \; ;$$

$$\frac{1}{Oy_3^{\,2}} = \frac{(Si\gamma)^2}{a^4} + \frac{(Sj\gamma)^2}{b^4} + \frac{(Sk\gamma)^2}{c^4} \; ;$$

$$\therefore \quad \frac{1}{Oy_1^{\,2}} + \frac{1}{Oy_2^{\,2}} + \frac{1}{Oy_3^{\,2}} = \frac{1}{a^4}\left\{(Sia)^2 + (Si\beta)^2 + (Si\gamma)^2\right\} + \&c.$$

$$= \frac{1}{a^2} + \frac{1}{b^2} + \frac{1}{c^2} \quad \text{(Ex. 7)}.$$

Ex. 9. *If through a fixed point within an ellipsoid three chords be drawn mutually at right angles, the sum of the reciprocals of the products of their segments will be constant.*

Let θ be the vector to the given point; a, β, γ unit vectors parallel to three chords at right angles to each other.

Then $\theta + xa = \rho$ gives

$$S(\theta + xa)\,\phi(\theta + xa) = 1$$

a quadratic equation in x, the product of whose roots is

$$\frac{S\theta\phi\theta - 1}{Sa\phi a} \; ;$$

$\therefore$ the product of the reciprocals of the segments of the chord is

$$\frac{1}{x_1 a x_2 a} = \frac{Sa\phi a}{S\theta\phi\theta - 1} \cdot \frac{1}{(Ta)^2} \; ;$$

and the sum of the reciprocals of the products of the segments is

$$\frac{1}{S\theta\phi\theta - 1} \cdot \left\{ \frac{Sa\phi a}{(Ta)^2} + \frac{S\beta\phi\beta}{(T\beta)^2} + \frac{S\gamma\phi\gamma}{(T\gamma)^2} \right\}.$$

Now since $\quad Sa\phi a = \dfrac{(Sia)^2}{a^2} + \dfrac{(Sja)^2}{b^2} + \dfrac{(Ska)^2}{c^2} \quad$ (63. 2, 3),

the sum of the reciprocals of the products

$$= \frac{1}{S\theta\phi\theta - 1} \left[\frac{1}{a^2}\left\{(Sia)^2 + (Si\beta)^2 + (Si\gamma)^2\right\} \right.$$

$$\left. + \frac{1}{b^2}\left\{(Sja)^2 + (Sj\beta)^2 + (Sj\gamma)^2\right\} \right.$$

$$+ \frac{1}{c^3} \left\{ (Ska)^2 + \dots\dots\dots\dots \right\} \Big]$$

$$= \frac{1}{S\theta\phi\theta - 1} \left(\frac{1}{a^2} + \frac{1}{b^2} + \frac{1}{c^2} \right) \quad (31. \text{ Cor.}).$$

Cor. If θ be not constant, but $S\theta\phi\theta$ be so, i. e. if the given point be situated on an ellipsoid concentric with and similar to the given ellipsoid, the same is true.

Ex. 10. *If the poles lie in a plane parallel to yz, the polar planes cut the axis of x always in the same point.*

Let pi be the distance from the origin of the plane in which the poles lie, δ any line in that plane, then $\pi = pi + \delta$ is the vector to a pole, and

$$S\rho\phi (pi + \delta) = 1 \quad (59)$$

the equation of the corresponding polar plane.

At the point where this plane cuts the axis of x,

$$\rho = xi;$$
$$\therefore \ Spxi\phi i + xSi\phi\delta = 1.$$

Now δ is a vector in a plane perpendicular to ϕi,

$$\therefore \ Si\phi\delta = S\delta\phi i = 0;$$

and $\qquad Si\phi i = \text{constant} = n$ suppose;

$$\therefore \ npx = 1,$$

which shews that x is constant.

Ex. 11. *A, B and C are three similar and similarly situated ellipsoids; A and B are concentric, and C has its centre on the surface of B. To shew that the tangent plane to B at this point is parallel to the plane of intersection of A and C.*

Let a be the vector to the centre of C.

$$S\rho\phi\rho = a \quad \text{the equation of } A,$$
$$S\rho\phi\rho = b \quad \dots\dots\dots\dots\dots B,$$
$$S (\rho - a) \phi (\rho - a) = c \dots\dots C.$$

Now at the intersection of A and C, ρ is the same for both; therefore the equation of the plane of intersection is to be found by subtracting the one from the other.

It is therefore $2S\rho\phi a = Sa\phi a + a - c$;

and the equation of the tangent plane to B at the centre of C is

$$S\pi\phi a = b \; ;$$

∴ both planes are perpendicular to ϕa, and are consequently parallel.

Ex. 12. *If through a given point chords be drawn to an ellipsoid, the intersections of pairs of tangent planes at their extremities all lie in a plane parallel to the tangent plane at the extremity of the diameter which passes through the point.*

Let a be the vector to the point; $a + x_1\beta$, $a + x_2\beta$, the vectors to the points of intersection with the ellipsoid of chords parallel to β ; then

$$S\pi\phi \left(a + x_1\beta\right) = 1,$$
$$S\pi\phi \left(a + x_2\beta\right) = 1,$$

are the equations of the tangent planes at these points.

At the intersection of these planes π is the same for both; ∴ subtracting we get

$$S\pi\phi\beta = 0,$$
$$S\pi\phi a = 1.$$

The last equation is that of the line of intersection of the tangent planes; and that line is perpendicular to ϕa, or (57. Cor.) parallel to the tangent plane at the extremity of the diameter which passes through the given point.

Cor. $S\pi\phi\beta = 0$ shews that the line of intersection corresponding to any one chord is parallel to the tangent plane at the extremity of the diameter which is parallel to that chord.

Ex. 13. *Two similar and similarly situated ellipsoids are cut by a series of ellipsoids similar and similarly situated to the two given ones; and in such a manner that the planes of intersection are at right angles to one another. Shew that the centres of the cutting ellipsoids lie on another ellipsoid.*

Let $$S\rho\phi\rho = 1 \dots\dots\dots\dots\dots\dots\dots(1),$$
$$S(\rho - a)\,\phi\,(\rho - a) = C \dots\dots\dots\dots\dots(2),$$

be the given ellipsoids;
$$S(\rho - \pi)\,\phi\,(\rho - \pi) = x\dots\dots\dots\dots\dots(3),$$

one of the cutting ellipsoids.

ϕ is the same for all because the ellipsoids are similar.

The plane of intersection of (1) and (3) is found by subtracting the equations; and is therefore
$$2S\rho\phi\pi = S\pi\phi\pi + 1 - x.$$

The plane of intersection of (2) and (3) is
$$2S\rho\,(\phi\pi - \phi a) = S\pi\phi\pi - Sa\phi a + C - x.$$

The former of these planes is perpendicular to $\phi\pi$ and the latter to $\phi\pi - \phi a$; and, since by the question, the former is perpendicular to the latter, $\phi\pi$ is perpendicular to $\phi\pi - \phi a$,
$$\therefore\ S\phi\pi\,(\phi\pi - \phi a) = 0,$$

the equation of the locus of the centres of the cutting ellipsoids.

This equation will be reduced to the requisite form by observing that
$$S\phi\pi\phi\pi = S\pi\phi\phi\pi = S\pi\phi^2\pi$$
$$S\phi\pi\phi a = Sa\phi^2\pi\,;$$
$$\therefore\ S(\pi - a)\,\phi^2\pi = 0,$$

the equation of an ellipsoid of which the semi-axes are proportional to
$$a^2,\ b^2,\ c^2\quad (63.\ 6).$$

The Cartesian equation is
$$\frac{x^2}{a^4} + \frac{y^2}{b^4} + \frac{z^2}{c^4} - \left(\frac{xx'}{a^4} + \frac{yy'}{b^4} + \frac{zz'}{c^4}\right) = 0.$$

Ex. 14. *If a tangent plane be drawn to the inner of two similar concentric and similarly situated ellipsoids the point of contact is the centre of the elliptic section of the outer ellipsoid.*

Let $S\rho\phi\rho = 1$ be the equation of the inner,

$a^2 S\rho\phi\rho = 1$ of the outer ellipsoid.

The tangent plane is $S\pi\phi\rho = 1$.

Now if σ be the vector to the elliptic section measured from the point of contact, $\pi = \rho + \sigma$ is a point in the outer ellipsoid;

$$\therefore a^2 S(\rho + \sigma)\phi(\rho + \sigma) = 1.$$

But $S\sigma\phi\rho = 0$ (57. Cor.);

$$\therefore a^2 + a^2 S\sigma\phi\sigma = 1,$$

$$\frac{a^2}{1-a^2} S\sigma\phi\sigma = 1,$$

the equation of an ellipse of which the centre is the point of contact.

Ex. 15. *Find the equation of the curve described by a given point in a line of given length whose extremities move in fixed straight lines.*

First, let the straight lines lie in one plane.

Let unit vectors parallel to them be a, β.

Let the vectors of the extremities of the moving line be xa, $y\beta$, and its length l. Then the condition is

$$(y\beta - xa)^2 = - l^2,$$

or $$x^2 + y^2 + 2xy Sa\beta = l^2 \quad (1).$$

The vector to a point which divides this line in the ratio $e : 1$ is

$$\rho = xa + e(y\beta - xa)$$
$$= xa(1 - e) + ey\beta;$$
$$\therefore Sa\rho = -(1 - e)x + ey Sa\beta,$$
$$S\beta\rho = (1 - e)x Sa\beta - ey;$$

whence $\qquad x = \dfrac{Sa\rho + Sa\beta S\beta\rho}{(1-e)\,(S^2 a\beta - 1)}, \quad y = \dfrac{S\beta\rho + Sa\beta Sa\rho}{e\,(S^2 a\beta - 1)}$,

which values being substituted in equation (1) give the required equation, viz. :

$$\frac{(Sa\rho + Sa\beta S\beta\rho)^2}{(1-e)^2} + \frac{(S\beta\rho + Sa\beta Sa\rho)^2}{e^2}$$

$$+\, 2\,\frac{Sa\beta}{e\,(1-e)}\,(Sa\rho + Sa\beta S\beta\rho)\,(S\beta\rho + Sa\beta Sa\rho)$$

$$= l^2\,(S^2 a\beta - 1)^2.$$

But ρ is subject to the additional condition (31. 2. Cor. 2) $S\,.\,a\beta\rho = 0$; and the locus is a plane ellipse.

When the given straight lines are at right angles to one another, the equation is much simplified, for

$$Sa\beta = 0\,;$$

and our equations are

$$x^2 + y^2 = l^2,$$

$$Sa\rho = -\,(1-e)\,x, \quad S\beta\rho = -\,ey\,;$$

whence $\qquad \dfrac{(Sa\rho)^2}{(1-e)^2} + \dfrac{(S\beta\rho)^2}{e^2} = l^2,$

an ellipse of which the semi-axes are le and $l\,(1-e)$.

Generally, if the given lines do not meet, let the origin be chosen midway along the line perpendicular to both; then we have

$$\{\gamma + xa - (-\gamma + y\beta)\}^2 = -\,l^2,$$

γ and $-\gamma$ being the vectors perpendicular to the lines,

$$\rho = (\gamma + xa)\,(1-e) + e\,(-\gamma + y\beta).$$

The first gives

$$4\gamma^2 + (xa - y\beta)^2 = -\,l^2\,;$$

and the second gives, as in the simpler case above,

$$Sa\rho = -\,(1-e)\,x + eySa\beta,$$

$$S\beta\rho = (1-e)\,xSa\beta - ey.$$

Hence the elimination of x and y again leads to the equation of an ellipsoid, the only difference being that l^2 is diminished by the square of the shortest distance between the lines; i.e. the axes are less than in the former case.

In the extreme case, where $l = 2T\gamma$, the equation cannot be satisfied except by
$$x = 0, \quad y = 0,$$
(i.e. the locus is reduced to a single point), unless indeed we have
$$a = \pm \beta,$$
for then $x = \pm y,$

and the locus is a straight line parallel to each of the preceding lines.

65. *The cone.*

1. To find the equation of a cone of revolution whose vertex is the origin O.

Let a be a unit vector along the axis OA,

     ρ the vector to a point P on the surface of the cone;

then $Sa\rho = -T\rho \cos \theta,$

θ being the angle POA.

But this angle is constant,

       $\therefore \ S^2 a\rho = c^2 \rho^2$ is the equation required.

2. The equation of a cone which has circular sections, but which is not necessarily a cone of revolution, is thus found.

Take the vertex as the origin, and let one of the circular sections be the intersection of the plane
$$Sa\rho = -a^2 \dots\dots\dots\dots\dots\dots(1)$$
with the sphere $\rho^2 = S\beta\rho \dots\dots\dots\dots\dots\dots(2).$

Since these are scalar equations we may multiply them together; and thus obtain at all the points of the circular section
$$a^2 \rho^2 + Sa\rho S\beta\rho = 0 \dots\dots\dots\dots\dots\dots(3).$$

Now if $x\rho$ or ρ' be written in place of ρ, the equation is not changed, since ρ occurs twice on each side. It is therefore the required equation of the cone.

COR. 1. Every section by a plane parallel to $Sa\rho = -a^2$ is a circle.

For the equation of a plane parallel to
$$Sa\rho = -a^2$$
is
$$Sa\rho = -aa^2,$$

which being substituted in the equation of the cone gives
$$\rho^2 = aS\beta\rho,$$

the equation of a circle.

COR. 2. The plane $\qquad S\beta\rho = -b\beta^2 \dots\dots\dots\dots\dots\dots\dots(4)$
also gives a circle whose equation is
$$a^2\rho^2 = b\beta^2 Sa\rho \dots\dots\dots\dots\dots\dots\dots(5).$$

These two equations give the *subcontrary* sections.

To deduce the relation between the two sections; let O be the vertex of the cone, OAB the plane through a, β; AB the line in which the section cuts this plane, AD that in which the subcontrary section cuts it;
$$OA = \rho, \quad OB = \rho', \quad OD = x\rho'.$$
We have, by (5), $\qquad x\rho'^2 = \dfrac{b\beta^2}{a^2} Sa\rho'$
$$= -b\beta^2, \text{ by (1)},$$
$$= S\beta\rho, \text{ by (4)},$$
$$= \rho^2, \text{ by (2)};$$
i.e. $\qquad\qquad\qquad OB \cdot OD = OA^2,$

and the triangles OAB, OAD are similar, or AD cuts OA at the same angle that AB cuts OB.

66. If $\qquad \phi\rho = 2a^2\rho + aS\beta\rho + \beta Sa\rho,$
the equation of the cone is reduced to
$$S\rho\phi\rho = 0.$$

It is evident that all the properties of $\phi\rho$, Art. 44, are applicable here.

As in Art. 57, the equation of the tangent plane is
$$S\pi\phi\rho = 0.$$

67.   EXAMPLES.

Ex. 1.   *Tangent planes are drawn to an ellipsoid from a given external point, to find the cone which has its vertex at the origin [the centre of the ellipsoid], and which passes through all the points of contact of the tangent planes with the ellipsoid.*

Let α be the vector to the external point, ρ a point in the ellipsoid where a tangent plane through α touches it.

Then the equation of the ellipsoid is
$$S\rho\phi\rho = 1,$$
and the equation of the tangent plane
$$S\alpha\phi\rho = 1, \quad \text{i. e. } S\rho\phi\alpha = 1.$$

The equation
$$S\rho\phi\rho = (S\rho\phi\alpha)^2,$$
or
$$\frac{x^2}{a^2} + \frac{y^2}{b^2} + \frac{z^2}{c^2} = \left(\frac{xx'}{a^2} + \frac{yy'}{b^2} + \frac{zz'}{c^2}\right)^2,$$

represents a surface passing through the points of contact; and is the cone required.   [For it is homogeneous in $T\rho$.]

Ex. 2.   *Of a system of three rectangular vectors two are confined to given planes, to find the surface traced out by the third.*

Let π, ρ, σ be the three vectors, of which two are confined to given planes whose equations are
$$S\alpha\pi = 0, \quad S\beta\rho = 0,$$
to find the locus of σ.

Since the vectors are at right angles, we have
$$S\pi\rho = 0, \quad S\pi\sigma = 0, \quad S\sigma\rho = 0,$$
and we have five equations from which to eliminate π and ρ.

Since $$S\alpha\pi = 0, \quad S\sigma\pi = 0,$$
π is at right angles to both α and σ, and therefore to the plane $\alpha\sigma$; or
$$\pi = x\,V\alpha\sigma.$$

Since $\qquad S\beta\rho = 0,\quad S\sigma\rho = 0,$

ρ is at right angles to the plane $\beta\sigma$; therefore

$$\rho = yV\beta\sigma,$$

and $\qquad\qquad\qquad \pi\rho = xyVa\sigma V\beta\sigma.$

Now $\qquad\qquad\qquad\qquad S\pi\rho = 0,$

therefore $\qquad\qquad\qquad S \,.\, Va\sigma V\beta\sigma = 0,$

or $\qquad\qquad\quad S\,(a\sigma - Sa\sigma)\,(\beta\sigma - S\beta\sigma) = 0,$

or $\qquad\qquad\qquad \sigma^2 Sa\beta - Sa\sigma S\beta\sigma = 0,$

the equation of a cone of the second order, which has circular sections (65. 2).

Cor. The circular sections are parallel to the two planes to which the two vectors are confined.

Ex. 3. *The equation* $\rho = t^2 a + u^2\beta + (t+u)^2\gamma$ *is that of a cone of the second order touched by each of the three planes through* $OAB,\ OBC,\ OCA$; *and the section* ABC *through the extremities of* $a,\ \beta,\ \gamma$ *is an ellipse touched at their middle points by* $AB,\ BC,\ CA.$

1. If the surface be referred to oblique co-ordinates parallel to $a,\ \beta,\ \gamma$ respectively, we shall have

$$\rho = xa + y\beta + z\gamma,$$

therefore $\qquad\quad x = t^2,\quad y = u^2,\quad z = (t+u)^2,$

or $\qquad\qquad z = (\sqrt{x} + \sqrt{y})^2 = x + y + 2\sqrt{xy},$

which gives $\qquad\qquad (z - x - y)^2 = 4xy,$

a cone of the second order.

2. If $t = -u$, the equation becomes

$$\rho = t^2\,(a+\beta),$$

the equation of a straight line bisecting the base AB, which since it satisfies the equation relative to t, shews that this line coincides with the cone in all its length; i.e. the cone is touched in this line by the plane OAB.

Similarly, by putting $t = 0,\ u = 0$ respectively, we can shew that the cone is touched by the plane $BOC,\ COA$ in the lines which bisect $AC,\ CA.$

$$10\text{—}2$$

3. Restricting ourselves to the plane ABC, we have the section of a cone of the second order enclosed by the triangle ABC, which triangle is itself the section of three planes each of which touches the cone.

Ex. 4. *The equation* $\rho = a\alpha + b\beta + c\gamma$ *with the condition* $ab + bc + ca = 0$ *is a cone of the second order, and the lines* OA, OB, OC *coincide throughout their length with the surface.*

1. It is evident that the equation gives

$$xy + yz + zx = 0.$$

2. That if $b = 0$, $c = 0$, the question is satisfied by

$$\rho = a\alpha,$$

whatever be a, therefore &c.

Ex. 5. *Find the locus of a point, the sum of the squares of whose distances from a number of given planes is constant.*

Let $S\delta_1\rho_1 = C_1$, $S\delta_2\rho_2 = C_2$, &c. be the equations of the given planes, ρ the vector to the point under consideration; then $x_1\delta_1$, $x_2\delta_2$, &c. will be the perpendiculars on the planes from the point; provided

$$\rho + x_1\delta_1 = \rho_1, \quad \rho + x_2\delta_2 = \rho_2, \quad \&c.;$$

therefore $S\delta_1 (\rho + x_1\delta_1) = C_1$, &c.

and $x_1\delta_1^2 = C_1 - S\delta_1\rho$, &c.,

$$x_1^2\delta_1^4 = (C_1 - S\delta_1\rho)^2;$$

i.e. the square of the line perpendicular to the first plane from the given point

$$= \left(\frac{C_1 - S\delta_1\rho}{T\delta_1}\right)^2,$$

and, by the question,

$$\left(\frac{C_1 - S\delta_1\rho}{T\delta_1}\right)^2 + \left(\frac{C_2 - S\delta_2\rho}{T\delta_2}\right)^2 + \&c. \text{ is constant.}$$

The locus is therefore a surface of the second order.

Ex. 6. *The lines which divide proportionally the pairs of opposite sides of a gauche quadrilateral, are the generating lines of a hyperbolic paraboloid.*

Let $ABCD$ be the quadrilateral.

AD, BC are divided proportionally in P and R.

Let $CA = a$, $CB = \beta$, $CD = \gamma$;

$\qquad CR = m\beta$, $DP = mDA$;

i. e. $\qquad CP - \gamma = m(a - \gamma)$;

therefore $\qquad RP = CP - CR = \gamma + m(a - \gamma) - m\beta$,

$$\rho = CQ = CR + pRP$$
$$= m\beta + p\{\gamma + m(a - \gamma) - m\beta\}$$
$$= xa + y\beta + z\gamma, \text{ say};$$

therefore $\qquad x = pm, \quad y = m - pm, \quad z = p(1 - m)$;

therefore $\qquad m = x + y, \quad p = \dfrac{x}{x + y}$,

$$z = \frac{x}{x + y} - x,$$

or $\qquad\qquad (x + z)(x + y) = x,$

the equation referred to oblique co-ordinates parallel to a, β, γ.

PASCAL'S HEXAGRAM.

68. Let O be the origin, OA, OB, OC, OD, OE five given vectors lying on the surface of a cone, and terminated in a plane section of the cone $ABCDEF$, not passing through O; OX *any* vector lying on the same surface.

Let $OA = a$, $OB = \beta$, $OC = \gamma$, $OD = \delta$, $OE = \epsilon$, $OX = \rho$.

The equation

$$S \cdot V(Va\beta V\delta\epsilon)\, V(V\beta\gamma V\epsilon\rho)\, V(V\gamma\delta V\rho a) = 0 \ldots\ldots\ldots(1)$$

is the equation of a cone of the second order whose vertex is O and vector ρ along the surface.   For

1. It is a cone whose vertex is O because it is not altered by writing $x\rho$ for ρ. Also it is of the second order in ρ, since ρ occurs in it twice and twice only.

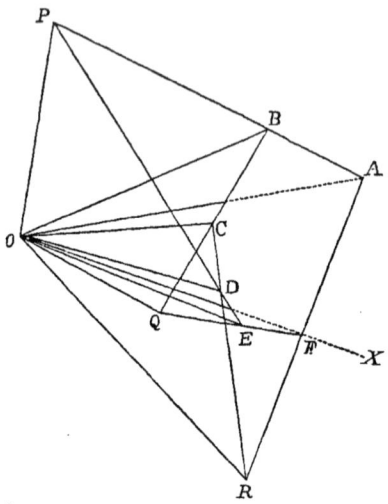

2. All the vectors OA, OB, OC, OD, OE lie on its surface.

This we shall prove by shewing that if ρ coincide with any one of them the equation (1) is satisfied.

If ρ coincide with α, the last term of the left-hand side of the equation, viz. $V\rho\alpha$, becomes $V\alpha\alpha = V\alpha^2 = 0$, and the equation is satisfied.

If ρ coincide with β, the left-hand side of the equation becomes

$$S \cdot V(V\alpha\beta V\delta\epsilon)\ V(V\beta\gamma V\epsilon\beta)\ V(V\gamma\delta V\beta\alpha)\ldots\ldots\ldots(2).$$

Now $V(V\beta\gamma V\epsilon\beta) = -V(V\epsilon\beta V\beta\gamma)$, (22. 2), is a vector parallel to β (31. 3), call it $m\beta$; and

$$V.\{V(V\alpha\beta V\delta\epsilon)\ V(V\gamma\delta V\beta\alpha)\} = V.\{V(V\alpha\beta V\delta\epsilon)\ V(V\alpha\beta V\gamma\delta)\},\ (22.\ 2),$$
$$= \text{a multiple of } V\alpha\beta,\ (31.\ 3),$$
$$= nV\alpha\beta,\text{ say.}$$

Hence the product of the first and third vectors in expression (2) becomes

$$\text{scalar} + n\,V\alpha\beta,$$

and the second is $m\beta$; therefore expression (2) becomes, by 31. 2,

$$S \cdot (\text{scalar} + nV\alpha\beta)\, m\beta$$
$$= mnS\beta V\alpha\beta$$
$$= 0,$$

because $V\alpha\beta$ is a vector perpendicular to β.

Equation (1) is therefore satisfied when ρ coincides with β.

If ρ coincide with γ both the second and third vectors are parallel to β (31. 3); therefore their product is a scalar, and equation (1) is satisfied.

The other cases are but repetitions of these.

Hence equation (1) is satisfied if ρ coincide with any one of the five vectors α, β, γ, δ, ϵ; i.e. OA, OB, OC, OD, OE are vectors on the surface of the cone.

3.   Let F be the point in which OX cuts the plane $ABCDE$; then $ABCDEF$ are the angular points of a hexagon inscribed in a conic section.

4.   Let the planes OAB, ODE intersect in OP; OBC, OEF in OQ; OCD, OFA in OR; then

$$V \cdot V\alpha\beta V\delta\epsilon = mOP, \; (31.\ 4),$$
$$V \cdot V\beta\gamma V\epsilon\rho = nOQ,$$
$$V \cdot V\gamma\delta V\rho\alpha = pOR;$$

therefore

$$S \cdot V\,(V\alpha\beta V\delta\epsilon)\ V\,(V\beta\gamma V\epsilon\rho)\ V\,(V\gamma\delta V\rho\alpha) = mnpS(OP \cdot OQ \cdot OR);$$

hence equation (1) gives

$$S\,(OP \cdot OQ \cdot OR) = 0,$$

or (31. 2. Cor. 2) OP, OQ, OR are in the same plane.

Hence PQR, the intersection of this plane with the plane $ABCDEF$ is a straight line. But P is the point of intersection of AB, ED, &c.

Therefore, the opposite sides (1st and 4th, 2nd and 5th, 3rd and 6th) of a hexagon inscribed in a conic section being produced .meet in the same straight line.

Cor. It is evident that the demonstration applies to any six points in the conic, whether the lines which join them form a hexagon or not.

ADDITIONAL EXAMPLES TO CHAP. VIII.

1. Find the locus of a point, the ratio of whose distances from two given straight lines is constant.

2. Find the locus of a point the square of whose distance from a given line is proportional to its distance from a given plane.

3. Prove that the locus of the foot of the perpendicular from the centre on the tangent plane of an ellipsoid is

$$(ax)^2 + (by)^2 + (cz)^2 = (x^2 + y^2 + z^2)^2.$$

4. The sum of the squares of the reciprocals of any three radii at right angles to one another is constant.

5. If Oy_1, Oy_2, Oy_3 be perpendiculars from the centre on tangent planes at the extremities of conjugate diameters, and if Q_1, Q_2, Q_3 be the points where they meet the ellipsoid; then

$$\frac{1}{OY_1^2 . OQ_1^2} + \frac{1}{OY_2^2 . OQ_2^2} + \frac{1}{OY_3^2 . OQ_3^2} = \frac{1}{a^4} + \frac{1}{b^4} + \frac{1}{c^4}.$$

6. If tangent planes to an ellipsoid be drawn from points in a plane parallel to that of xy, the curves which contain all the points of contact will lie in planes which all cut the axis of z in the same point.

7. Two similar and similarly situated ellipsoids intersect in a plane curve whose plane is conjugate to the line which joins the centres of the ellipsoids.

8. If points be taken in conjugate semi-diameters produced, at distances from the centre equal to p times those semi-diameters respectively; the sum of the squares of the reciprocals of the

perpendiculars from the centre on their polar planes is equal to p^2 times the sum of the squares of the perpendiculars from the centre on tangent planes at the extremities of those diameters.

9. If P be a point on the surface of an ellipsoid, PA, PB, PC any three chords at right angles to each other, the plane ABC will pass through a fixed point, which is in the normal to the ellipsoid at P; and distant from P by

$$\frac{\dfrac{2}{p}}{\dfrac{1}{a^2} + \dfrac{1}{b^2} + \dfrac{1}{c^2}},$$

where p is the perpendicular from the centre on the tangent plane at P.

10. Find the equation of the cone which has its vertex in a given point, and which touches and envelopes a given ellipsoid.

CHAPTER IX.

FORMULÆ AND THEIR APPLICATION.

69. PRODUCTS of two or more vectors.

1. *Two vectors.* The relations which exist between the scalars and vectors of the product of two vectors have already been exhibited in Art. 22. We simply extract them :

(a) $Sa\beta = S\beta a.$ (b) $Va\beta = -V\beta a.$

(c) $a\beta + \beta a = 2Sa\beta.$ (d) $a\beta - \beta a = 2Va\beta.$

These we shall quote as formulæ (1).

2. We may here add a single conclusion for quaternion products.

Any quaternion, such as $a\beta$, may be written as the sum of a scalar and a vector. If therefore q and r be quaternions, we may write

$$q = Sq + Vq,$$
$$r = Sr + Vr;$$

therefore $\quad qr = SqSr + SqVr + SrVq + VqVr,$

and $\quad S.qr = SqSr + S.VqVr,$

$$V.qr = SqVr + SrVq + V.VqVr,$$

where $S.VqVr$ is the scalar part, and $V.VqVr$ the vector part of the product of the two vectors Vq, Vr.

If now we transpose q and r, and apply (a) and (b) of formulæ 1, we get

$$\left. \begin{array}{l} S.qr = S.rq \\ V.qr + V.rq = 2\left(SqVr + SrVq\right) \end{array} \right\} \dots\dots\dots\dots(2).$$

3. *Three vectors.* By observing that $S.\gamma S\alpha\beta$ is simply the scalar of a vector, and is consequently zero, we may insert or omit such an expression at pleasure. By bearing this in mind the reader will readily apprehend the demonstrations which follow, even in cases where we have studied brevity.

$$S.\alpha\beta\gamma = S.(S\alpha\beta+V\alpha\beta)\gamma$$
$$= S.\gamma V\alpha\beta, \text{ (by 1. } a),$$
$$= S.\gamma(S\alpha\beta+V\alpha\beta)$$
$$= S.\gamma\alpha\beta \dots\dots\dots\dots\dots\dots\dots(3).$$

Again, $$S.\alpha\beta\gamma = S.\alpha(S\beta\gamma+V\beta\gamma)$$
$$= S(V\beta\gamma.\alpha), \text{ (by 1. } a),$$
$$= S(S\beta\gamma+V\beta\gamma)\alpha$$
$$= S.\beta\gamma\alpha \dots\dots\dots\dots\dots\dots\dots(3).$$

The formulæ marked (3) shew that a change of order amongst three vectors produces no change in the scalar of their product, provided the cyclical order remain unchanged.

This conclusion might have been obtained by a different process, thus:

In (2) let $q=\alpha\beta$, $r=\gamma$, there results at once
$$S.\alpha\beta\gamma = S.\gamma\alpha\beta.$$

Again in (2) let $q=\gamma\alpha$, $r=\beta$, there results
$$S.\gamma\alpha\beta = S.\beta\gamma\alpha.$$

We have therefore, as before,
$$S.\alpha\beta\gamma = S.\gamma\alpha\beta = S.\beta\gamma\alpha \dots\dots\dots\dots\dots(3).$$

4. $$S.\alpha\beta\gamma = S.\alpha V\beta\gamma$$
$$= -S.\alpha V\gamma\beta, \text{ (by 1. } b),$$
$$= -S.\alpha\gamma\beta \dots\dots\dots\dots\dots\dots\dots(4).$$

Similarly $$S.\alpha\beta\gamma = -S.\beta\alpha\gamma \dots\dots\dots\dots\dots\dots(4),$$

or a cyclical change of order amongst three vectors changes the sign of the scalar of their product.

5.   It has already been seen (Art. 31. 1) that $-S . \alpha\beta\gamma$ is the volume of the parallelepiped of which the three edges which terminate in the point O are the lines OA, OB, OC whose vectors are α, β, γ respectively.

We may express this volume in the form of a determinant, thus :

Let α, β, γ be replaced by

$$xi + yj + zk, \quad x'i + y'j + z'k, \quad x''i + y''j + z''k \text{ (Art. 31. 5) ;}$$

x, y, z being the rectangular co-ordinates of A, x', y', z' those of B, x'', y'', z'' those of C, measured from O as the origin ; then

$$S . \alpha\beta\gamma = S . (xi \ + yj \ + zk)$$
$$\times (x'i \ + y'j \ + z'k)$$
$$\times (x''i + y''j + z''k).$$

Now if we observe first that the scalar part of this product is confined to those terms in which all the three vectors i, j, k appear; and secondly that the sign of any term in the product will by formulæ (3) and (4) be $-$ or $+$ according as cyclical order is or is not retained, we perceive that we have the exact conditions which apply to a determinant : therefore

$$S . \alpha\beta\gamma = - \begin{vmatrix} x , & y , & z \\ x', & y', & z' \\ x'', & y'', & z'' \end{vmatrix} \quad \ldots\ldots\ldots\ldots\ldots(5).$$

The volume of the pyramid $OABC$ is one-sixth of the above.

Note relative to the sign of the scalar.

Since $ijk = -1$ (19), it is clear that if OA, OB, OC assume the positions of Ox, Oy, Oz in the figure of Art. 16, $S (OA . OB . OC)$ will have a *minus* sign, whilst the order of the letters A, B, C is right-handed as seen from O.

If now we take any pyramid whatever $OABC$, of which the vertex is O, and assume that $S (OA . OB . OC)$ (which, being proportional to the volume of the pyramid, we may designate $OABC$), is negative when the order of the letters A, B, C is right-handed as seen from O, we shall find the following general law of signs to hold good whatever be the vertex; viz. *the sign of the scalar is minus or plus according as the order in it of the angles of the base of the ·pyramid is right-handed or left-handed as seen from the vertex.*

For example,
$$CABO = S\,(CA\,.\,CB\,.\,CO)$$
$$= S\,(\alpha - \gamma)\,(\beta - \gamma)\,(-\gamma)$$
$$= -\,S\alpha\beta\gamma$$
$$= -\,OABC,$$

which is *plus* because $OABC$ is *minus,* and the order of the letters A, B, O as seen from C is left-handed.

6.
$$V.\,\alpha\beta\gamma = V\,.\,\alpha\,(S\beta\gamma + V\beta\gamma)$$
$$= \alpha S\beta\gamma + V\,.\,\alpha V\beta\gamma\,;$$
$$V.\,\gamma\beta\alpha = V\,.\,(S\gamma\beta + V\gamma\beta)\,\alpha$$
$$= \alpha S\beta\gamma - V.\,\alpha V\gamma\beta,\ (1.\ b),$$
$$= \alpha S\beta\gamma + V.\,\alpha V\beta\gamma,\ (1.\ b),$$
$$= V.\,\alpha\beta\gamma \ \dots\dots\dots\dots\dots\dots\dots\dots\dots(6).$$

7.
$$V.\,\alpha\beta\gamma = V\,.\,(S\alpha\beta + V\alpha\beta)\,\gamma$$
$$= \gamma S\alpha\beta - V.\,\gamma V\alpha\beta\,;$$
$$V.\,\gamma\alpha\beta = V\,.\,\gamma\,(S\alpha\beta + V\alpha\beta)$$
$$= \gamma S\alpha\beta + V.\,\gamma V\alpha\beta\,;$$

therefore $V.\,\alpha\beta\gamma + V.\,\gamma\alpha\beta = 2\gamma S\alpha\beta \dots\dots\dots\dots\dots\dots(7).$

8. $2V.\,\alpha V\beta\gamma = V.\,\alpha\,(\beta\gamma - \gamma\beta),\ (1.\ d),$
$$= V.\,\alpha\beta\gamma + V.\,\gamma\alpha\beta - (V.\,\alpha\gamma\beta + V.\,\gamma\alpha\beta)$$
$$= V(\alpha\beta\gamma + \beta\alpha\gamma) - V\,(\alpha\gamma\beta + \gamma\alpha\beta),\ \text{(by 6),}$$
$$= V.\,(\alpha\beta + \beta\alpha)\,\gamma - V.\,(\alpha\gamma + \gamma\alpha)\,\beta$$
$$= 2\gamma S\alpha\beta - 2\beta S\alpha\gamma,\ (1.\ c)\,;$$

therefore $V.\,\alpha V\beta\gamma = \gamma S\alpha\beta - \beta S\alpha\gamma \dots\dots\dots\dots\dots\dots(8).$

9. We have, by (8),

$$V \cdot a V \beta \gamma = \gamma S a \beta - \beta S a \gamma,$$
$$V \cdot \beta V \gamma a = a S \beta \gamma - \gamma S a \beta,$$
$$V \cdot \gamma V a \beta = \beta S a \gamma - a S \beta \gamma;$$

therefore, by addition,

$$V \cdot (a V \beta \gamma + \beta V \gamma a + \gamma V a \beta) = 0 \quad \ldots \ldots \ldots \ldots (9).$$

10. $$V \cdot a \beta \gamma = V \cdot a (S \beta \gamma + V \beta \gamma)$$
$$= a S \beta \gamma + V \cdot a V \beta \gamma,$$

which, by (8), $$= a S \beta \gamma - \beta S a \gamma + \gamma S a \beta \ldots \ldots \ldots (10).$$

Another proof of this important formula is found in the identity

$$\frac{1}{2} (a \beta \gamma + \gamma \beta a) = \frac{1}{2} a (\beta \gamma + \gamma \beta) - \frac{1}{2} \beta (a \gamma + \gamma a) + \frac{1}{2} \gamma (a \beta + \beta a),$$

which, by (4) and (6), is the theorem itself.

11. If in (8) we write $V a \beta$ in place of a, we get

$$V \cdot V a \beta V \beta \gamma = \gamma S (V a \beta \cdot \beta) - \beta S (V a \beta \cdot \gamma)$$
$$= \gamma S \cdot a \beta \beta - \beta S a \beta \gamma$$
$$= - \beta S \cdot a \beta \gamma \quad \ldots \ldots \ldots \ldots \ldots \ldots \ldots (11).$$

12. *Four vectors.* If in (8) we write $V a \delta$ in place of a, we obtain

$$V (V a \delta V \beta \gamma) = \gamma S \cdot a \delta \beta - \beta S \cdot a \delta \gamma \ldots \ldots \ldots (12).$$

13. By (12) we have

$$V (V \beta \gamma V a \delta) = \delta S \cdot \beta \gamma a - a S \cdot \beta \gamma \delta.$$

But $$V (V \beta \gamma V a \delta) = - V (V a \delta V \beta \gamma).$$

Hence, by adding the above result to (12), we get

$$\delta S \cdot \beta \gamma a - a S \cdot \beta \gamma \delta + \gamma S \cdot a \delta \beta - \beta S \cdot a \delta \gamma = 0,$$

which, by (3) and (4), if we adopt alphabetical order, may be written

$$a S \cdot \beta \gamma \delta - \beta S \cdot a \gamma \delta + \gamma S \cdot a \beta \delta - \delta S \cdot a \beta \gamma = 0 \quad \ldots \ldots (13),$$

or $$\delta S \cdot a \beta \gamma = a S \cdot \beta \gamma \delta - \beta S \cdot a \gamma \delta + \gamma S \cdot a \beta \delta \quad \ldots \ldots \ldots (13),$$

or, again, if we adopt cyclical order,

$$aS . \beta\gamma\delta - \delta S . a\beta\gamma + \gamma S . \delta a\beta - \beta S . \gamma\delta a,$$

or, finally, $\delta S . a\beta\gamma = aS . \beta\gamma\delta - \beta S . \gamma\delta a + \gamma S . \delta a\beta$ ....(13).

This equation expresses a vector in terms of three other vectors. The following equation expresses it in terms of the vectors which result from their products two and two.

14. $V(\gamma\delta a\beta)$ may be written, first as $V(\gamma . \delta a\beta)$, and secondly as $V(\gamma\delta . a\beta)$, and the results compared. These forms give respectively

$$\begin{aligned}
V(\gamma . \delta a\beta) &= V . \gamma (S . \delta a\beta + V . \delta a\beta) \\
&= \gamma S . a\beta\delta + V . \gamma (\delta S a\beta - a S \delta\beta + \beta S \delta a), \text{ by (3) and (10),} \\
&= \gamma S . a\beta\delta + V\gamma\delta S a\beta - V\gamma a S \delta\beta + V\gamma\beta S \delta a ; \\
V(\gamma\delta . a\beta) &= V . (S\gamma\delta + V\gamma\delta) (S a\beta + V a\beta) \\
&= V a\beta S\gamma\delta + V\gamma\delta S a\beta + V . V\gamma\delta V a\beta \\
&= V a\beta S\gamma\delta + V\gamma\delta S a\beta - V . V a\beta V\gamma\delta \\
&= V a\beta S\gamma\delta + V\gamma\delta S a\beta - \delta S . a\beta\gamma + \gamma S . a\beta\delta, \text{ by (12).}
\end{aligned}$$

The two expressions being equated, and the common terms deleted, there results

$$\delta S . a\beta\gamma = V a\beta S\gamma\delta + V\beta\gamma S a\delta + V\gamma a S\beta\delta \ldots\ldots\ldots(14).$$

15. $\begin{aligned}[t]
S . a\beta\gamma\delta &= S . (S . a\beta\gamma + V . a\beta\gamma) \delta \\
&= S . (V . a\beta\gamma) \delta \\
&= S . (a S\beta\gamma - \beta S a\gamma + \gamma S a\beta) \delta, \text{ by (10),} \\
&= S a\beta S\gamma\delta - S a\gamma S\beta\delta + S a\delta S\beta\gamma \ldots\ldots\ldots\ldots(15).
\end{aligned}$

16. $\begin{aligned}[t]
S (V a\beta V\gamma\delta) &= S . (a\beta - S a\beta) (\gamma\delta - S\gamma\delta) \\
&= S . a\beta\gamma\delta - S a\beta S\gamma\delta \\
&= S a\delta S\beta\gamma - S a\gamma S\beta\delta, \text{ by (15)} \ldots\ldots\ldots(16).
\end{aligned}$

17. $\begin{aligned}[t]
S . a\beta\gamma\delta &= S . (V a\beta\gamma) \delta \\
&= S . \delta V a\beta\gamma \\
&= S . \delta a\beta\gamma \ldots\ldots\ldots\ldots\ldots\ldots\ldots(17).
\end{aligned}$

18. *Five vectors.* As we do not purpose to exhibit any applications of the relations which exist among five or more vectors, we shall confine ourselves to simply writing down the two following expressions.

$$S \cdot \alpha\beta\gamma\delta\epsilon = -S \cdot \epsilon\delta\gamma\beta\alpha,$$
$$V \cdot \alpha\beta\gamma\delta\epsilon = V \cdot \epsilon\delta\gamma\beta\alpha \dots\dots\dots\dots\dots(18).$$

70. Many of these formulæ might have been proved differently, and some of them more directly, by assuming for instance that α, β, γ are not in the same plane. In this case *any* other vector δ may be expressed in terms of α, β, γ, by the equation

$$\delta = x\alpha + y\beta + z\gamma, \ (31. \ 5);$$

therefore $S \cdot \beta\gamma\delta = xS \cdot \beta\gamma\alpha = xS \cdot \alpha\beta\gamma, \ (3),$

$\qquad\qquad S \cdot \gamma\delta\alpha = yS \cdot \gamma\beta\alpha = -yS \cdot \alpha\beta\gamma, \ (4),$

$\qquad\qquad S \cdot \delta\alpha\beta = zS \cdot \gamma\alpha\beta = zS \cdot \alpha\beta\gamma, \ (3);$

therefore $\delta S \cdot \alpha\beta\gamma = x\alpha S \cdot \alpha\beta\gamma + y\beta S \cdot \alpha\beta\gamma + z\gamma S \cdot \alpha\beta\gamma$

$\qquad\qquad\qquad = \alpha S \cdot \beta\gamma\delta - \beta S \cdot \gamma\delta\alpha + \gamma S \cdot \delta\alpha\beta$

which is formula 13.

71. EXAMPLES.

Ex. 1. *To express the relation between the sides of a spherical triangle and the angles opposite to them.*

Retaining the notation and figure of Ex. 2, Art. 29, we shall have

$$V\alpha\beta V\beta\gamma = \gamma' \sin c \cdot \alpha' \sin a,$$

where γ', α' are unit vectors perpendicular respectively to the planes *OAB*, *OBC*.

Therefore $V \cdot V\alpha\beta V\beta\gamma = \sin c \sin a \cdot \beta \sin B.$

Also $-\beta S \cdot \alpha\beta\gamma = \beta \sin c \sin \phi, \ (31. \ 1),$

where φ is the angle between *OC* and the plane *OAB*.

Now these results are equal (formula 11), therefore

$$\sin \phi = \sin a \sin B.$$

Similarly $\sin \phi = \sin b \sin A$;

therefore $\sin a \sin B = \sin b \sin A,$

or $\sin a \; : \; \sin b \; :: \; \sin A \; : \; \sin B.$

Ex. 2.   *To find the condition that the perpendiculars from the angles of a tetrahedron on the opposite faces shall intersect one another.*

Let OA, OB, OC be the edges of the tetrahedron (Fig. of Art. 31), a, β, γ the corresponding vectors.

Vector perpendiculars from A and B on the opposite faces are $V\beta\gamma$, $V\gamma a$ respectively (22. 8). If these perpendiculars intersect in G, the three points A, B, G will be in one plane, whence

$$S . (\beta - a) \; V\beta\gamma V\gamma a = 0 \;\; (31. \; 2, \; \text{Cor. } 2),$$

i.e. $S . (\beta - a) \; V . \; V\beta\gamma V\gamma a = 0.$

Now $V . \; V\beta\gamma V\gamma a = - \gamma S . \; \beta\gamma a$ (Formula 11),

therefore $S . (\beta - a) \; V . \; V\beta\gamma V\gamma a = - (S\beta\gamma - Sa\gamma) \, S . \, \beta\gamma a.$

Hence $S\beta\gamma = Sa\gamma.$

Now $BC^2 + OA^2 = (\gamma - \beta)^2 + a^2$

$$= a^2 + \beta^2 + \gamma^2 - 2S\beta\gamma$$
$$= a^2 + \beta^2 + \gamma^2 - 2Sa\gamma$$
$$= (\gamma - a)^2 + \beta^2$$
$$= AC^2 + OB^2.$$

Consequently the condition that all three perpendiculars shall meet in a point is that the sum of the squares of each pair of opposite edges shall be the same.

Cor.   Conversely, if the sum of the squares of each pair of opposite edges is the same, the perpendiculars from the angles on the opposite faces will meet in a point.

Ex. 3.   *If P be a point in the face ABC of a tetrahedron, from which are drawn Pa, Pb, Pc, respectively parallel to OA, OB, OC to meet the opposite faces OBC, OCA, OAB in a, b, c; then will*

$$\frac{Pa}{OA} + \frac{Pb}{OB} + \frac{Pc}{OC} = 1.$$

T. Q.   **11**

Retaining the notation of the last examples, let $OP = \delta$, $Pa = -x\alpha$, $Pb = -y\beta$, $Pc = -z\gamma$; then

$$Oa = \delta - x\alpha, \quad Ob = \delta - y\beta, \quad Oc = \delta - z\gamma.$$

Now because P, A, B, C are in the same plane

$$S . (\delta - \alpha)(\alpha - \beta)(\beta - \gamma) = 0,$$

i.e. $$S . \delta(\alpha\beta + \beta\gamma + \gamma\alpha) = S . \alpha\beta\gamma \dots\dots\dots\dots(1);$$

and because O, a, B, C are in the same plane

$$S . (\delta - x\alpha)\beta\gamma = 0,$$

i.e. $$xS\alpha\beta\gamma = S . \delta\beta\gamma \dots\dots\dots\dots\dots(2);$$

also because O, A, b, C are in the same plane

$$S . (\delta - y\beta)\gamma\alpha = 0,$$

i.e. $$yS . \beta\gamma\alpha = S . \delta\gamma\alpha,$$

or, by formula 3, $$yS . \alpha\beta\gamma = S . \delta\gamma\alpha \dots\dots\dots\dots\dots(3);$$

lastly, because O, A, B, c are in the same plane

$$S . (\delta - z\gamma)\alpha\beta = 0,$$

i.e. $$zS . \gamma\alpha\beta = S . \delta\alpha\beta,$$

or $$zS . \alpha\beta\gamma = S . \delta\alpha\beta \dots\dots\dots\dots\dots(4).$$

Adding (2), (3), and (4) there results

$$(x + y + z) S . \alpha\beta\gamma = S . \delta\beta\gamma + S . \delta\gamma\alpha + S . \delta\alpha\beta$$
$$= S . \alpha\beta\gamma, \text{ by } (1),$$

therefore $$x + y + z = 1:$$

hence $$\frac{Pa}{OA} + \frac{Pb}{OB} + \frac{Pc}{OC} = 1.$$

Cor. 1. If P be in the plane ABC produced below the plane OBC, Pa as a vector will have the same sign as OA has; hence in this case we shall have

$$-\frac{Pa}{OA} + \frac{Pb}{OB} + \frac{Pc}{OC} = 1.$$

Cor. 2.   If P be outside both the planes OBC, OCA; we shall have

$$-\frac{Pa}{OA} - \frac{Pb}{OB} + \frac{Pc}{OC} = 1.$$

Ex. 4.   *Any point Q is joined to the angular points A, B, C, O of a tetrahedron, and the joining lines, produced if necessary, meet the opposite faces in a, b, c, o; to prove that*

$$\frac{Qa}{Aa} + \frac{Qb}{Bb} + \frac{Qc}{Cc} + \frac{Qo}{Oo} = 1;$$

regard being had to the signs of Aa, Bb, &c., as in the last example.

Let $QA = \alpha$, $QB = \beta$, $QC = \gamma$, $QO = \delta$; $Qa = a\alpha$, $Qb = b\beta$, $Qc = c\gamma$, $Qo = d\delta$: then since the points a, b, c, o are in the planes BCO, ACO, ABO, ABC, respectively, we have, as in the last example,

$$aS \cdot \alpha(\beta\gamma + \gamma\delta + \delta\beta) = S \cdot \beta\gamma\delta,$$

$$\text{&c.} \qquad\qquad \text{&c.}$$

i.e.

$$aS \cdot (\alpha\beta\gamma + \alpha\gamma\delta + \alpha\delta\beta) - S \cdot \beta\gamma\delta = 0 \ldots\ldots\ldots\ldots (1),$$

$$bS \cdot (\beta\alpha\gamma + \beta\gamma\delta + \beta\delta\alpha) - S \cdot \alpha\gamma\delta = 0 \ldots\ldots\ldots\ldots (2),$$

$$cS \cdot (\gamma\alpha\beta + \gamma\beta\delta + \gamma\delta\alpha) - S \cdot \alpha\beta\delta = 0 \ldots\ldots\ldots\ldots (3),$$

$$dS \cdot (\delta\alpha\beta + \delta\beta\gamma + \delta\gamma\alpha) - S \cdot \alpha\beta\gamma = 0 \ldots\ldots \ldots\ldots\ldots (4).$$

Now, if we write

$$S \cdot \alpha\beta\gamma = x, \quad S \cdot \alpha\gamma\delta = y, \quad S \cdot \alpha\delta\beta = z, \quad S \cdot \beta\gamma\delta = u;$$

and apply the formulæ 3 and 4, we get

$$ax + ay + az - u = 0,$$

$$-bx - y - bz + bu = 0,$$

$$cx + cy + z - cu = 0,$$

$$- x - dy - dz + du = 0,$$

which give

$$\frac{a}{a-1}x + \frac{d}{d-1}u = 0,$$

$$\frac{a}{a-1}y + \frac{b}{b-1}u = 0,$$

11—2

$$\frac{c}{c-1}y - \frac{b}{b-1}z = 0,$$

$$\frac{c}{c-1}x - \frac{d}{d-1}z = 0;$$

and, therefore, $\quad \dfrac{1}{a-1} + \dfrac{b}{b-1} + \dfrac{c}{c-1} + \dfrac{d}{d-1} = 0,$

i. e. $\quad\quad\quad \dfrac{a}{a-1} + \dfrac{b}{b-1} + \dfrac{c}{c-1} + \dfrac{d}{d-1} = 1,$

or $\quad\quad\quad\quad \dfrac{Qa}{Aa} + \dfrac{Qb}{Bb} + \dfrac{Qc}{Cc} + \dfrac{Qo}{Oo} = 1.$

Ex. 5. *If two tetrahedra ABCD, A'B'C'D' are so situated that the straight lines, AA', BB', CC', DD' all meet in a point, the lines of intersection of the planes of corresponding faces shall all lie in the same plane.*

Let $A'A$, $B'B$, $C'C$, $D'D$ meet in O.

$$OA = a, \quad OB = \beta, \quad OC = \gamma, \quad OD = \delta,$$

$$OA' = ma, \quad OB' = n\beta, \quad OC' = p\gamma, \quad OD' = q\delta.$$

The equation of the plane ABC is (34. 5)

$$S\rho\,(V a\beta + V\beta\gamma + V\gamma a) = S\,.\,a\beta\gamma,$$

and that of $A'B'C'$ becomes, after dividing both sides by mnp,

$$S\rho\left(\frac{1}{p}V a\beta + \frac{1}{m}V\beta\gamma + \frac{1}{n}V\gamma a\right) = S\,.\,a\beta\gamma.$$

The vector line of intersection of the two planes is (34. 9)

$$V\,.\,(V a\beta + V\beta\gamma + V\gamma a)\left(\frac{1}{p}V a\beta + \frac{1}{m}V\beta\gamma + \frac{1}{n}V\gamma a\right),$$

i.e. by formula (11), omitting the common factor $S\,.\,a\beta\gamma$,

$$\left(\frac{1}{n} - \frac{1}{p}\right)a + \left(\frac{1}{p} - \frac{1}{m}\right)\beta + \left(\frac{1}{m} - \frac{1}{n}\right)\gamma.$$

From this expression the vectors of the intersections of the other planes may at once be written down.

That of ABD, $A'B'D'$ is

$$\left(\frac{1}{n} - \frac{1}{q}\right)\alpha + \left(\frac{1}{q} - \frac{1}{m}\right)\beta + \left(\frac{1}{m} - \frac{1}{n}\right)\delta\,;$$

that of ACD, $A'C'D'$ is

$$\left(\frac{1}{p} - \frac{1}{q}\right)\alpha + \left(\frac{1}{q} - \frac{1}{m}\right)\gamma + \left(\frac{1}{m} - \frac{1}{p}\right)\delta\,;$$

and that of BCD, $B'C'D'$

$$\left(\frac{1}{p} - \frac{1}{q}\right)\beta + \left(\frac{1}{q} - \frac{1}{n}\right)\gamma + \left(\frac{1}{n} - \frac{1}{p}\right)\delta.$$

Now to prove that any three of these lines lie in the same plane, all that is necessary is to prove (31. 2, Cor. 2) that the scalar of the product of their vectors equals 0.

If we take the vectors of the first three, we may write them under the form

$$a\alpha + b\beta + c\gamma, \quad a'\alpha + b'\beta + c\delta, \quad a''\alpha + b'\gamma - b\delta,$$

respectively; so that the scalar of their product is

$$S\,.\,(a\alpha + b\beta + c\gamma)\,(a'\alpha + b'\beta + c\delta)\,(a''\alpha + b'\gamma - b\delta).$$

Now the coefficient of every different scalar in this product is separately equal to 0. That of $S\,.\,\alpha\beta\gamma$ for instance is, omitting the common factor b',

$$\left(\frac{1}{n} - \frac{1}{p}\right)\left(\frac{1}{q} - \frac{1}{m}\right) - \left(\frac{1}{m} - \frac{1}{n}\right)\left(\frac{1}{p} - \frac{1}{q}\right) - \left(\frac{1}{p} - \frac{1}{m}\right)\left(\frac{1}{n} - \frac{1}{q}\right),$$

in which every term vanishes.

That again of $S\,.\,\beta\gamma\delta$ is

$$-bcb' + cb'b,$$

which is 0; and so of the rest.

Hence the intersections, two and two, of the first three pairs of planes lie in the same plane; and the same may be proved in like manner of any other three : whence the truth of the proposition.

Ex. 6. *CP, CD are conjugate semi-diameters of an ellipse, as also CP', CD' ; PP', DD' are joined; to prove that the area of the triangle PCP' equals that of the triangle DCD'.*

Let a, β, a', β' be the vectors CP, CD, CP', CD'; k a unit vector perpendicular to the plane of the ellipse.

Since
$$a = \psi^{-1}\psi a = -(aiSi\psi a + bjSj\psi a), \&c., \&c. \ (47.5),$$

therefore $Vaa' = V.(aiSi\psi a + bjSj\psi a)(aiSi\psi a' + bjSj\psi a')$

$$= abk (Si\psi a Sj\psi a' - Sj\psi a Si\psi a')$$

$$= -abkS.kV(\psi a\psi a'). \quad (\text{Formula 16.})$$

Similarly $V\beta\beta' = -abkS.kV(\psi\beta\psi\beta')$.

Now ψa, $\psi\beta$ are unit vectors at right angles to one another; as are also $\psi a'$, $\psi\beta'$; therefore the angle between ψa and $\psi a'$ is the same as that between $\psi\beta$ and $\psi\beta'$.

Hence $S.kV(\psi a\psi a') = S.kV(\psi\beta\psi\beta')$,

and $Vaa' = V\beta\beta'$,

i.e. area of triangle PCP' = that of triangle DCD'. ·

Ex. 7. *If a parallelepiped be constructed on the semi-conjugate diameters of an ellipsoid, the sum of the squares of the areas of the faces of the parallelepiped is equal to the sum of the squares of the faces of the rectangular parallelepiped constructed on the semi-axes.*

By 63. 9, $a = -(aiSi\psi a + bjSj\psi a + ckSk\psi a)$

$$\beta = -(aiSi\psi\beta + bjSj\psi\beta + ckSk\psi\beta);$$

therefore $V a\beta = abk (Si\psi a Sj\psi\beta - Si\psi\beta Sj\psi a)$

$$+ acj (Si\psi a Sk\psi\beta - Si\psi\beta Sk\psi a)$$

$$+ bci (Sj\psi a Sk\psi\beta - Sj\psi\beta Sk\psi a).$$

Now $Si\psi a Sj\psi\beta - Si\psi\beta Sj\psi a = SVij V\psi\beta\psi a$, Formula (16),

$$= -Sk\psi\gamma, \ (\text{Art. 17});$$

therefore
$$V\alpha\beta = -(abkSk\psi\gamma + acjSj\psi\gamma + bciSi\psi\gamma),$$
$$V\gamma\alpha = -(abkSk\psi\beta + acjSj\psi\beta + bciSi\psi\beta),$$
$$V\beta\gamma = -(abkSk\psi\alpha + acjSj\psi\alpha + bciSi\psi\alpha).$$

If now we square and add these expressions, observing that because $\psi\alpha$, $\psi\beta$, $\psi\gamma$ are unit vectors at right angles to one another,
$$(Si\psi\alpha)^2 + (Si\psi\beta)^2 + (Si\psi\gamma)^2 = 1,$$

we shall have
$$(V\alpha\beta)^2 + (V\alpha\gamma)^2 + (V\beta\gamma)^2 = -\{(ab)^2 + (ac)^2 + (bc)^2\},$$

which (21. 4) is the proposition to be proved.

Ex. 8. *To find the locus of the intersections of tangent planes at the extremities of conjugate diameters of an ellipsoid.*

Let π be the vector to the point of intersection of tangent planes at the extremities of α, β, γ: then
$$S\pi\phi\alpha = 1, \quad (57),$$

gives
$$S\pi\psi^2\alpha = -1,$$

or
$$S\psi\pi\psi\alpha = -1,$$
$$S\psi\pi\psi\beta = -1,$$
$$S\psi\pi\psi\gamma = -1.$$

From these three equations we extricate $\psi\pi$ by means of formula (14), which gives
$$\psi\pi S\psi\alpha\psi\beta\psi\gamma = V\psi\alpha\psi\beta S\psi\pi\psi\gamma + V\psi\beta\psi\gamma S\psi\pi\psi\alpha$$
$$+ V\psi\gamma\psi\alpha S\psi\pi\psi\beta;$$

therefore
$$\psi\pi = V\psi\alpha\psi\beta + V\psi\beta\psi\gamma + V\psi\gamma\psi\alpha$$
$$= \psi\gamma + \psi\alpha + \psi\beta,$$
$$(\psi\pi)^2 = -(1 + 1 + 1)$$
$$= -3,$$
$$\frac{x^2}{3a^2} + \frac{y^2}{3b^2} + \frac{z^2}{3c^2} = 1;$$

an ellipsoid similar to the given ellipsoid.

Ex. 9. *If O, A, B, C, D, E are any six points in space, OX any given direction, OA', OB', OC', OD', OE' the projections of OA, OB, OC, OD, OE on OX; $BCDE$, $CDEA$, $DEAB$, $EABC$, $ABCD$ the volumes of the pyramids whose vertices are B, C, D, E, A, with a positive or negative sign in accordance with the law given in the note to 69. 5; then*

$$OA'. BCDE + OB'. CDEA + OC'. DEAB + OD'. EABC$$
$$+ OE'. ABCD = 0.$$

Let OA, OB, OC, OD, OE be α, β, γ, δ, ϵ respectively.

Write for $\alpha S (\gamma - \beta)(\delta - \beta)(\epsilon - \beta)$ its value

$$\alpha (S. \gamma\delta\epsilon - S. \delta\epsilon\beta + S. \epsilon\beta\gamma - S. \beta\gamma\delta),$$

and similar expressions for $\beta S (\alpha - \gamma)(\delta - \gamma)(\epsilon - \gamma)$, &c., and there will result, by addition,

$$\alpha S(\gamma - \beta)(\delta - \beta)(\epsilon - \beta) + \beta S (\alpha - \gamma)(\delta - \gamma)(\epsilon - \gamma)$$
$$+ \gamma S (\alpha - \delta)(\beta - \delta)(\epsilon - \delta) + \delta S (\alpha - \epsilon)(\beta - \epsilon)(\gamma - \epsilon)$$
$$+ \epsilon S (\beta - \alpha)(\gamma - \alpha)(\delta - \alpha) = 0,$$

i.e. retaining the notation adopted in the Note referred to,

$$OA. BCDE + OB. CDEA + OC. DEAB + OD. EABC$$
$$+ OE. ABCD = 0.$$

Now let π be a vector along OX; then the operation by $S. \pi$ on the above expression gives the result required.

———————

In some of the examples which follow, we will endeavour to shew how a problem should *not*, as well as how it should, be attacked.

Ex. 10. *Given any three planes, and the direction of the vector perpendicular to a fourth, to find its length so that they may meet in one point.*

Let $S\alpha\rho = a$, $S\beta\rho = b$, $S\gamma\rho = c$ be the three, and let δ be the vector perpendicular to the new plane. Then, if its equation be

$$S\delta\rho = d,$$

we must find the value of d that these four equations may all be satisfied by one value of ρ.

Formula (14) gives

$$\rho S \cdot a\beta\gamma = Va\beta S\gamma\rho + V\beta\gamma Sa\rho + V\gamma a S\beta\rho$$
$$= cVa\beta + aV\beta\gamma + bV\gamma a,$$

by the equations of the first three. Operate by $S \cdot \delta$, and use the fourth equation, and we have the required value

$$dS \cdot a\beta\gamma = aS \cdot \beta\gamma\delta + bS \cdot \gamma a\delta + cS \cdot a\beta\delta.$$

Ex. 11. *The sum of the (vector) areas of the faces of any tetrahedron, and therefore of any polyhedron, is zero.*

Take one corner as origin, and let a, β, γ be the vectors of the other three. Then the vector areas of the three faces meeting in the origin are

$$\frac{1}{2}Va\beta, \quad \frac{1}{2}V\beta\gamma, \quad \frac{1}{2}V\gamma a, \text{ respectively.}$$

That of the fourth may be expressed in any of the forms

$$\frac{1}{2}V(\gamma - a)(\beta - a), \quad \frac{1}{2}V(a - \beta)(\gamma - \beta), \quad \frac{1}{2}V(\beta - \gamma)(a - \gamma).$$

But all of these have the common value

$$\frac{1}{2}V(\gamma\beta + \beta a + a\gamma),$$

which is obviously the sum of the three other vector-areas taken negatively. Hence the proposition, which is an elementary one in Hydrostatics.

Now any polyhedron may be cut up by planes into tetrahedra, and the faces exposed by such treatment have vector-areas equal and opposite in sign. Hence the extension.

Ex. 12. *If the pressure be uniform throughout a fluid mass, an immersed tetrahedron (and therefore any polyhedron) experiences no couple tending to make it rotate.*

This is supplementary to the last example. The pressures on the faces are fully expressed by the vector-areas above given, and their points of application are the centres of inertia of the areas of the faces. The co-ordinates of these points are

$$\frac{1}{3}(a+\beta), \quad \frac{1}{3}(\beta+\gamma), \quad \frac{1}{3}(\gamma+a), \quad \frac{1}{3}(a+\beta+\gamma),$$

and the sum of the couples is

$$\frac{1}{6}V.\{Va\beta.(a+\beta)+V\beta\gamma.(\beta+\gamma)+V\gamma a.(\gamma+a)$$
$$+V(\gamma\beta+\beta a+a\gamma).(a+\beta+\gamma)\}$$
$$=-\frac{1}{6}V(Va\beta.\gamma+V\beta\gamma.a+V\gamma a.\beta)=0,$$

by applying formula (9).

Ex. 13. *What are the conditions that the three planes*

$$Sa\rho=a, \quad S\beta\rho=b, \quad S\gamma\rho=c,$$

shall intersect in a straight line?

There are many ways of attacking such a question, so we will give a few for practice.

(a) $$\rho S.a\beta\gamma = Va\beta S\gamma\rho + V\beta\gamma Sa\rho + V\gamma a S\beta\rho$$
$$= cVa\beta + aV\beta\gamma + bV\gamma a$$

by the given equations. But this gives a single definite value of ρ unless both sides vanish, so that the conditions are

$$S.a\beta\gamma = 0,$$

and $$cVa\beta + aV\beta\gamma + bV\gamma a = 0,$$

which *includes* the preceding.

(b) $$S(la-m\beta)\rho = al-bm$$

is the equation of any plane passing through the intersection of the first two given planes. Hence, if the three intersect in a straight line there must be values of l, m such that

$$la-m\beta=\gamma,$$
$$la-mb=c.$$

The first of these gives, as before,

$$S.a\beta\gamma=0,$$

and it also gives

$$V\gamma a = m V a\beta, \quad V\beta\gamma = -l V a\beta,$$

so that if we multiply the second by $V a\beta$,

$$la V a\beta - mb V a\beta = c V a\beta$$

becomes $$-a V\beta\gamma - b V\gamma a = c V a\beta \, ;$$

the second condition of (a).

(c)   Again, suppose ρ to be given by the first two in the form

$$\rho = pa + q\beta + x V a\beta,$$

we find $$a = pa^2 + qSa\beta, \text{ because } SaVa\beta = 0,$$

$$b = pSa\beta + q\beta^2 \, ;$$

therefore

$$\rho \begin{vmatrix} a^2, & Sa\beta \\ Sa\beta, & \beta^2 \end{vmatrix} = a \begin{vmatrix} a, & Sa\beta \\ b, & \beta^2 \end{vmatrix} + \beta \begin{vmatrix} a^2, & a \\ Sa\beta, & b \end{vmatrix} + x V a\beta,$$

so that the third equation gives, operating by $S \cdot \gamma$,

$$c \begin{vmatrix} a^2, & Sa\beta \\ Sa\beta, & \beta^2 \end{vmatrix} = Sa\gamma \begin{vmatrix} a, & Sa\beta \\ b, & \beta^2 \end{vmatrix} + S\beta\gamma \begin{vmatrix} a^2, & a \\ Sa\beta, & b \end{vmatrix} + xS \cdot a\beta\gamma.$$

Now a determinate value of x would mean intersection in one point only ; so, as before,

$$S \cdot a\beta\gamma = 0,$$

$$c (a^2\beta^2 - S^2a\beta) = a (\beta^2 Sa\gamma - Sa\beta S\beta\gamma) - b (Sa\beta Sa\gamma - a^2 S\beta\gamma).$$

The latter may be written

$$S \cdot a [c (a\beta^2 - \beta Sa\beta) - a (\gamma\beta^2 - \beta S\beta\gamma) - b (aS\beta\gamma - \gamma Sa\beta)] = 0.$$

Now $$S \cdot a (a\beta^2 - \beta Sa\beta) = Sa (\beta \cdot \beta a - \beta S\beta a)$$

$$= S \cdot a (\beta V\beta a)$$

$$= - S \cdot a (\beta V a\beta) = - S (a\beta V a\beta).$$

Similarly, $$S \cdot a (\gamma\beta^2 - \beta S\beta\gamma) = S (a\beta V\beta\gamma),$$

and $$S \cdot a (aS\beta\gamma - \gamma Sa\beta) = S \cdot a (V \cdot \beta V\gamma a), \text{ (formula 8),}$$

$$= S (a\beta V\gamma a).$$

The equation now becomes

$$S \cdot a\beta (c V a\beta + a V\beta\gamma + b V\gamma a) = 0.$$

Now since $S \cdot \alpha\beta\gamma = 0$, α, β, γ are vectors in the same plane; therefore γ may be written $m\alpha + n\beta$,

$$\text{and } cV\alpha\beta + aV\beta\gamma + bV\gamma\alpha$$

assumes the form $eV\alpha\beta$, which, unless $e = 0$, gives

$$S(\alpha\beta V\alpha\beta) = 0,$$

or $V\alpha\beta$ is in the same plane with α, β; but it is also perpendicular to the plane, which is absurd; therefore $e = 0$, or

$$cV\alpha\beta + aV\beta\gamma + bV\gamma\alpha = 0 ;$$

thus the third and prolix method leads to the same conclusion as the first.

Ex. 14. *Find the surface traced out by a straight line which remains always perpendicular to a given line while intersecting each of two fixed lines.*

Let the equations of the fixed lines be

$$\varpi = a + x\beta, \quad \varpi_1 = a_1 + x_1\beta_1.$$

Then if ρ be the vector of the new line in any position,

$$\rho = \varpi + y(\varpi_1 - \varpi)$$
$$= (1 - y)(a + x\beta) + y(a_1 + x_1\beta_1).$$

This is not, as yet, the equation required. For it involves essentially *three* independent constants, x, x_1, y; and may therefore in general be made to represent any point whatever of infinite space. The reader may easily see this if he reflects that two lines which are not parallel must appear, from every point of space, to intersect one another. We have still to introduce the condition that the new line is perpendicular to a fixed vector, γ suppose, which gives

$$S \cdot \gamma(\varpi_1 - \varpi) = 0 = S \cdot \gamma[(a_1 - a) + x_1\beta_1 - x\beta].$$

This gives x_1 in terms of x, so that there are now but two indeterminates in the equation for ρ, which therefore represents a surface, which, it is not difficult to see, is one of the second order.

Ex. 15. *Find the condition that the equation*

$$S \cdot \rho\phi\rho = 1$$

may represent a surface of revolution.

The expression $\phi\rho$ here stands for something more general than that employed in Chap. VIII. above, in fact it may be written

$$\phi\rho = \alpha S\alpha_1\rho + \beta S\beta_1\rho + \gamma S\gamma_1\rho,$$

where α, α_1, β, β_1, γ, γ_1 are any six vectors whatever. This will be more carefully examined in the next chapter.

If the surface be one of revolution then, since it is central and of the second degree, it is obvious that any sphere whose centre is at the origin will cut it in two equal circles in planes perpendicular to the axis, and that these will be equidistant from the origin. Hence, if r be the radius of one of these circles, ϵ the vector to its centre, ρ the vector to any point in its circumference, it is evident that we have the following equation,

$$S\rho\phi\rho - 1 - C\left(\rho^2 + r^2\right) = (S\epsilon\rho)^2 - e^2,$$

where C and e are constants. This, being an identity, gives

$$\left. \begin{aligned} 1 - e^2 + Cr^2 &= 0 \\ S\rho\phi\rho - C\rho^2 &= (S\epsilon\rho)^2 \end{aligned} \right\}.$$

The form of these equations shews that C is an absolute constant, while r and e are related to one another by the first; and the second gives

$$\phi\rho = C\rho + \epsilon S\epsilon\rho.$$

This shews simply that $S \cdot \epsilon\rho\phi\rho = 0,$

i. e. ϵ, ρ, and $\phi\rho$ are coplanar, i. e. all the normals pass through a given straight line; or that the expression

$$V\rho\phi\rho,$$

whatever be ρ, expresses always a vector parallel to a particular plane.

Ex. 16. *If three mutually perpendicular vectors be drawn from a point to a plane, the sum of the reciprocals of the squares of their lengths is independent of their directions.*

Let $\qquad\qquad S\epsilon\rho = 1$

be the equation of the plane, and let α, β, γ be any set of mutually perpendicular unit-vectors. Then, if $x\alpha$, $y\beta$, $z\gamma$ be points in the plane, we have

$$xS\alpha\epsilon = 1, \quad yS\beta\epsilon = 1, \quad zS\gamma\epsilon = 1,$$

whence $\quad -\epsilon = \alpha S\alpha\epsilon + \beta S\beta\epsilon + \gamma S\gamma\epsilon \,(63.\ 2) = \dfrac{\alpha}{x} + \dfrac{\beta}{y} + \dfrac{\gamma}{z}.$

Taking the tensor, we have

$$T\epsilon^2 = \frac{1}{x^2} + \frac{1}{y^2} + \frac{1}{z^2}.$$

Ex. 17. *Find the equation of the straight line which meets, at right angles, two given straight lines.*

Let $\qquad\qquad \varpi = \alpha + x\beta, \quad \varpi = \alpha_1 + x_1\beta_1,$

be the two lines ; then the equation of the required line must be of the form

$$\varpi = \alpha_2 + x_2 V\beta\beta_1,$$

and nothing is undetermined but α_2.

Since the first and third equations denote lines having one point in common, we have

$$S \,.\, \beta \, V\beta\beta_1 \,(\alpha - \alpha_2) = 0.$$

Similarly $\qquad S \,.\, \beta_1 V\beta\beta_1 \,(\alpha_1 - \alpha_2) = 0.$

Let $\qquad\qquad \alpha_2 = y\beta + y_1\beta_1$

(it is obviously superfluous to add a term in $V\beta\beta_1$), then

$$S \,.\, \alpha\beta V\beta\beta_1 = y_1 T^2 V\beta\beta_1,$$
$$S \,.\, \alpha_1\beta_1 V\beta\beta_1 = -y T^2 V\beta\beta_1,$$

and, finally,

$$\varpi = \frac{1}{T^2 V\beta\beta_1} (\beta_1 S \,.\, \alpha\beta V\beta\beta_1 - \beta S \,.\, \alpha_1\beta_1 V\beta\beta_1) + x_2 V\beta\beta_1.$$

Ex. 18. *If $T\rho = T\alpha = T\beta = 1$, and $S \,.\, \alpha\beta\rho = 0$, shew that*

$$S \,.\, U(\rho - \alpha) \, U(\rho - \beta) = \sqrt{\frac{1}{2}(1 - S\alpha\beta)}.$$

Interpret this theorem geometrically.

We have, from the given equations, the following, which are equivalent to them,

$$\left.\begin{array}{c} \rho^2 = \alpha^2 = \beta^2 = -1 \\ \rho = x\alpha + y\beta \end{array}\right\}.$$

Hence $\qquad -x^2 - y^2 + 2xy S\alpha\beta = -1,$

$$U(\rho - \alpha) = \frac{(x-1)\,\alpha + y\beta}{\sqrt{(x-1)^2 - 2\,(xy-y)\,S\alpha\beta + y^2}},$$

$$U(\rho - \beta) = \frac{x\alpha + (y-1)\,\beta}{\sqrt{x^2 - 2\,(xy-x)\,S\alpha\beta + (y-1)^2}},$$

$S \cdot U(\rho - \alpha)\,U(\rho - \beta)$

$$= \frac{-x\,(x-1) + [xy + (x-1)\,(y-1)]\,S\alpha\beta - y\,(y-1)}{\sqrt{x^2 + y^2 - 2x + 1 - 2\,(xy-y)\,S\alpha\beta}\ \sqrt{x^2 + y^2 - 2y + 1 - 2\,(xy-x)\,S\alpha\beta}}$$

$$= \frac{x + y - (x+y-1)\,S\alpha\beta - 1}{\sqrt{2 - 2x + 2y S\alpha\beta}\ \sqrt{2 - 2y + 2x S\alpha\beta}}$$

$$= \frac{(x+y-1)\,(1 - S\alpha\beta)}{2\sqrt{(1-x-y)\,(1-S\alpha\beta) + xy\,\{1 - (S\alpha\beta)^2\}}}$$

$$= \frac{x+y-1}{2}\sqrt{\frac{1 - S\alpha\beta}{1 - x - y + xy\,(1 + S\alpha\beta)}}$$

$$= \frac{x+y-1}{2}\sqrt{\frac{1 - S\alpha\beta}{1 - x - y + \frac{1}{2}\,(2xy + x^2 + y^2 - 1)}}$$

$$= \frac{x+y-1}{\sqrt{2}}\sqrt{\frac{1 - S\alpha\beta}{1 - 2\,(x+y) + x^2 + y^2 + 2xy}}$$

$$= \pm\sqrt{\frac{1}{2}\,(1 - S\alpha\beta)}.$$

Of course there are far simpler solutions. Thus, for instance, the given equations shew that ρ, α, β are radii of some unit *circle*. Hence the expression is the cosine of the supplement of the angle between two chords of a circle drawn from the same point in the circumference. This is obviously half the angle subtended at the centre by radii drawn to the other ends of the chords. The cosine of this angle is

$$- Sa\beta,$$

and therefore the cosine of its half is

$$\sqrt{\frac{1}{2}(1 - Sa\beta)}.$$

Ex. 19. *Find the relative position, at any instant, of two points, which are moving uniformly in straight lines.*

If a', β' be their vector velocities, t the time elapsed since their vectors were a, β, their relative vector is

$$\rho = a + ta' - \beta - t\beta'$$
$$= (a - \beta) + t(a' - \beta'),$$

so that relatively to one another the motion is rectilinear, and the vector velocity is

$$a' - \beta'.$$

To find the time at which the mutual distance is least.

Here we may write

$$\rho = \gamma + t\delta,$$
$$T\rho^2 = -\gamma^2 - 2tS\gamma\delta - t^2\delta^2$$
$$= \frac{(S\gamma\delta)^2}{\delta^2} - \gamma^2 - \delta^2\left(t + \frac{S\gamma\delta}{\delta^2}\right)^2.$$

As the last term is positive, this is least when it vanishes, i.e. when

$$t = -S.\gamma\delta^{-1}.$$

This gives

$$\rho = \gamma - \delta S\gamma\delta^{-1}$$
$$= \gamma V\delta^{-1}\gamma,$$

the vector perpendicular drawn to the relative path; as is, of course, self-evident.

Ex. 20. *Find the locus of a given point in a line of given length, when the extremities of the line move in circles in one plane.* (Watt's *Parallel Motion.*)

Let σ and τ be the vectors of the ends of the line, drawn from the centres a, β of the circles. Then if ρ be the vector of the required point

$$\rho = (a + \sigma)(1 - e) + e(\beta + \tau),$$

subject to the conditions

$$\{a + \sigma - (\beta + \tau)\}^2 = -l^2,$$

$$S\gamma\sigma = 0, \ S\gamma\tau = 0,$$

$$\sigma^2 = -a^2, \ \tau^2 = -b^2.$$

From these equations σ and τ must be eliminated. We leave the work to the reader. There is obviously an equation of condition

$$S \cdot \gamma(\beta - a) = 0.$$

Ex. 21. *Classify the curves represented by an equation of the form*

$$\rho = \frac{a + x\beta + x^2\gamma}{a + bx + cx^2},$$

where a, β, γ are given vectors, and a, b, c given scalars.

In the first place we remark that x^2 in the numerator merely adds a constant vector to the value of ρ, unless $c = 0$.

Thus, if c do not vanish, the equation may be written, with a change of a and β and in general a change of origin,

$$\rho = \frac{a + x\beta}{a + bx + cx^2}:$$

and this again, by change of x and of a and β, as

$$\rho = \frac{a + x\beta}{a + cx^2}.$$

It is obvious that this represents a plane curve.

Also
$$\frac{Sa\rho}{S\beta\rho} = \frac{a^2 + xSa\beta}{Sa\beta + x\beta^2}.$$

T. Q.   12

Hence both numerator and denominator of x are of the first degree in $Sa\rho$, $S\beta\rho$; and therefore

$$Sa\rho = \frac{a^2 + xSa\beta}{a + cx^2}$$

gives an equation of the third degree in ρ by the elimination of x.

When we have $\quad\quad\quad Sa\beta = 0,$

$$Sa\rho = \frac{a^2}{a + cx^2},$$

$$S\beta\rho = \frac{x\beta^2}{a + cx^2},$$

whence $\quad\quad\quad\quad\quad x = \frac{a^2 S\beta\rho}{\beta^2 Sa\rho},$

and $\quad\quad a\,(Sa\rho)^2 + c\,\frac{a^4}{\beta^4}\,(S\beta\rho)^2 = a^2 Sa\rho,$

a conic section.

If $c = 0$, then with a change of x, a, β, γ, the equation may be written

$$\rho = \frac{a}{x} + \beta + x\gamma,$$

a hyperbola—so long at least as b does not also vanish.

If b and c both vanish, the equation is obviously that of a parabola.

If a and b both vanish, whilst c has a real value, we have again a parabola.

If a vanish while b and c have real values, we have again a hyperbola.

Ex. 22. *Find the locus of a point at which a given finite straight line subtends a given angle.*

Take the middle point of the line as origin, and let $\pm a$ be the vectors of its ends.   At ρ it subtends an angle whose cosine is

$$-SU(\rho - a)\,U(\rho + a).$$

This, equated to a constant, gives the locus required.   We may write the equation

$$a^2 - \rho^2 = cT(\rho - a)\,T(\rho + a).$$

This is, obviously, a surface of the fourth order; a ring or tore formed by the rotation of a circle about a chord.   When $c = 0$, i.e. when the angle is a right angle, the two sheets of this surface close up into the sphere

$$\rho^2 = a^2.$$

A plane section (in the plane a, β (suppose) where $T\beta = Ta$ and $Sa\beta = 0$) gives

$$\rho = xa + y\beta,$$

$$\{a^2(1 - x^2) - y^2 a^2\}^2 = c^2 \{(x - 1)^2 + y^2\}\{(x + 1)^2 + y^2\}\,a^4,$$

or $$\{1 - (x^2 + y^2)\}^2 = c^2 \{(x^2 + y^2 + 1)^2 - 4x^2\},$$

or, finally, $$1 - (x^2 + y^2) = \pm \frac{2cy}{\sqrt{1 - c^2}},$$

which, of course, denotes two equal circles intersecting at the ends of the fixed line.

Ex. 23.   *A ray of light falls on a thin reflecting cylinder, shew that it is spread over a right cone.*

Let a be the ray, τ a normal to the cylinder, ρ a reflected ray, β the axis of the cylinder.

Then τ is perpendicular to β, or

$$S\beta\tau = 0 \ldots\ldots\ldots\ldots\ldots\ldots\ldots\ldots(1).$$

Again ρ and a make equal angles with τ, on opposite sides of it, in one plane; therefore

$$\rho \parallel \tau a\tau$$

or $$V . \tau a\tau\rho = 0 \ldots\ldots\ldots\ldots\ldots\ldots\ldots(2).$$

12—2

Eliminating τ between (1) and (2) we have

$$\frac{\rho^2}{a^2} = \left(\frac{S\beta\rho}{S a\beta}\right)^2$$

the equation of the right cone of which β is the axis, and a a side.

ADDITIONAL EXAMPLES TO CHAP. IX.

1. Prove that $S \cdot (a + \beta)(\beta + \gamma)(\gamma + a) = 2S \cdot a\beta\gamma$.

2. $S \cdot V a\beta V\beta\gamma V\gamma a = -(Sa\beta\gamma)^2$.

3. $S \cdot V(V a\beta V\beta\gamma) V(V\beta\gamma V\gamma a) V(V\gamma a V a\beta) = -(S \cdot a\beta\gamma)^4$.

4. $S(V\beta\gamma V\gamma a) = \gamma^2 Sa\beta - S\beta\gamma S\gamma a$.

5. $a^2\beta^2\gamma^2 = (V a\beta\gamma)^2 - (Sa\beta\gamma)^2$

6. $\qquad = a^2(S\beta\gamma)^2 + \beta^2(S\gamma a)^2 + \gamma^2(Sa\beta)^2 - (Sa\beta\gamma)^2$
$\qquad - 2Sa\beta S\beta\gamma S\gamma a.$

7. $S(\gamma V \cdot a\beta\gamma) = \gamma^2 Sa\beta$.

8. $(a\beta\gamma)^2 = a^2\beta^2\gamma^2 + 2a\beta\gamma S \cdot a\beta\gamma$.

9. $S(V a\beta\gamma V\beta\gamma a V\gamma a\beta) = 4Sa\beta S\beta\gamma S\gamma a S \cdot a\beta\gamma$.

10. The expression

$$V a\beta V\gamma\delta + V a\gamma V\delta\beta + V a\delta V\beta\gamma$$

denotes a vector. What vector?

(Tait's *Quaternions.* Miscellaneous Ex. 1.)

11. $Sa\rho S \cdot \beta\gamma\delta - S\beta\rho S \cdot \gamma\delta a + S\gamma\rho S \cdot \delta a\beta - S\delta\rho S \cdot a\beta\gamma = 0$.

12. $(a\beta\gamma)^2 = 2a^2\beta^2\gamma^2 + a^2(\beta\gamma)^2 + \beta^2(a\gamma)^2 + \gamma^2(a\beta)^2 - 4a\gamma Sa\beta S\beta\gamma$.

(Hamilton, *Elements*, p. 346.)

13. With the notation of the Note, Art. 69. 5, we shall have

$$DABC = OABC - OBCD + OCDA - ODAB.$$

14. When A, B, C, D are in the same plane,

$$a \cdot BCD - \beta \cdot CDA + \gamma \cdot DAB - \delta \cdot ABC = 0,$$

where BCD, &c. are the areas of the triangles.

15. $\delta V \cdot a\beta\gamma + aV \cdot \beta\gamma\delta + \beta V \cdot \gamma\delta a + \gamma V \cdot \delta a\beta = 4S \cdot a\beta\gamma\delta.$

16. $Va\beta V\gamma\delta + V\beta\gamma V\delta a + V\gamma\delta Va\beta + V\delta a V\beta\gamma$ is a scalar. What is its geometrical meaning?

17. Find the equation of the sphere circumscribing a given tetrahedron.

18. A straight line intersects a fixed line at right angles, and turns uniformly about it while it slides uniformly along it. Find the equation of the surface described (1) when the fixed line is straight, (2) when it is circular.

CHAPTER X.

WITH the object of giving the student an idea of one of the physical applications of Quaternions, we will treat the solution of linear and vector equations from an elementary kinematical point of view. For this purpose we choose the problem of the deformation of a solid or fluid body, when all its parts are similarly and equally deformed.

DEF. *Homogeneous Strain* is such that portions of a body, originally equal, similar, and similarly placed, remain after the strain equal, similar, and similarly placed.

Thus straight lines remain straight lines, parallel lines remain parallel, equal parallel lines remain equal, planes remain planes, parallel planes remain parallel, and equal areas on parallel planes remain equal. Also the volumes of *all* portions of the body are increased or diminished in the same proportion, as is easily seen by supposing the body originally divided into small equal cubes by series of planes perpendicular to each other. After the strain, these cubes are all changed into similar, similarly placed, and equal parallelepipeds.

It is thus obvious that a homogeneous strain is entirely determined if we know into what vectors three given (non-coplanar) vectors are changed by it. Thus if α, β, γ become α', β', γ'

respectively: any other vector, which may of course be expressed as

$$\rho = \frac{1}{S \cdot \alpha\beta\gamma} (\alpha S \cdot \beta\gamma\rho + \beta S \cdot \gamma\alpha\rho + \gamma S \cdot \alpha\beta\rho)$$

is changed to

$$\rho' = \frac{1}{S \cdot \alpha\beta\gamma} (\alpha' S \cdot \beta\gamma\rho + \beta' S \cdot \gamma\alpha\rho + \gamma' S \cdot \alpha\beta\rho).$$

No needful generality is lost, while much simplification is gained, by taking α, β, γ as unit vectors at right angles to one another. This is, in fact, the method already spoken of, i.e. the imaginary division of the body into small equal cubes, by three mutually perpendicular series of equidistant planes. We thus have

$$\rho = -(\alpha S\alpha\rho + \beta S\beta\rho + \gamma S\gamma\rho),$$
$$\rho' = -(\alpha' S\alpha\rho + \beta' S\beta\rho + \gamma' S\gamma\rho).$$

Comparing these expressions we see that *Homogeneous Strain* alters a vector into a definite linear and vector function of its original value.

In abbreviated notation, we may write (as in Art. 63, though our symbol, as will soon be seen, is more general than that there employed)

$$\phi\rho = -(\alpha' S\alpha\rho + \beta' S\beta\rho + \gamma' S\gamma\rho),$$

where ϕ itself depends upon *nine* independent constants involved in the three equations

$$\left.\begin{array}{l} \phi\alpha = \alpha' \\ \phi\beta = \beta' \\ \phi\gamma = \gamma' \end{array}\right\}.$$

For α', β', γ' may of course be expressed in terms of α, β, γ: and, as they are quite independent of one another, the nine coefficients in the following equations may have absolutely any values whatever;

$$\left.\begin{array}{l} \phi\alpha = \alpha' = A\alpha + c\beta + b'\gamma \\ \phi\beta = \beta' = c'\alpha + B\beta + a\gamma \\ \phi\gamma = \gamma' = b\alpha + a'\beta + C\gamma \end{array}\right\} \dots\dots\dots\dots\dots\dots(a).$$

In discussing the particular form of ϕ which occurs in the treatment of central surfaces of the second order we found, Art. 44, that it possessed the property

$$S \cdot \sigma\phi\rho = S \cdot \rho\phi\sigma \dots\dots\dots\dots\dots\dots\dots(b),$$

whatever vectors are represented by ρ and σ. Remembering that a, β, γ form a rectangular unit system, we find from (a)

$$\begin{aligned} S \cdot \beta\phi a &= -c \\ S \cdot a\phi\beta &= -c' \end{aligned} \Big\},$$

with other similar pairs; so that our new value of ϕ satisfies (b) if, and only if, we have in (a)

$$\begin{aligned} a &= a' \\ b &= b' \\ c &= c' \end{aligned} \Bigg\} \dots\dots\dots\dots\dots\dots\dots\dots\dots(c).$$

The physical meaning of this condition, as will be seen immediately, is that the distortion expressed by ϕ takes place *without rotation*. In this case the nine constants are reduced to six.

But, although (b) is not generally true, we have

$$\begin{aligned} S \cdot \sigma\phi\rho &= -(Sa'\sigma Sa\rho + S\beta'\sigma S\beta\rho + S\gamma'\sigma S\gamma\rho) \\ &= -S \cdot \rho (aSa'\sigma + \beta S\beta'\sigma + \gamma S\gamma'\sigma), \end{aligned}$$

where the expression in brackets is a linear and vector function of σ, depending upon the same *nine* scalars as those in ϕ; and which we may therefore express by ϕ', so that

$$\phi'\sigma = -(aSa'\sigma + \beta S\beta'\sigma + \gamma S\gamma'\sigma)\dots\dots\dots\dots(d).$$

And with this we have obviously

$$S \cdot \sigma\phi\rho = S \cdot \rho\phi'\sigma\dots\dots\dots\dots\dots\dots\dots(e),$$

which is the general relation, of which (b) is a mere particular case.

By putting a, β, γ in succession for σ in (d) and referring to (a) we have

$$\begin{aligned} \phi'a &= Aa + c'\beta + b\gamma \\ \phi'\beta &= ca + B\beta + a'\gamma \\ \phi'\gamma &= b'a + a\beta + C\gamma \end{aligned} \Bigg\} \dots\dots\dots\dots\dots(f).$$

Comparing (f) with (a) we see that

$$\phi\rho = \phi'\rho,$$

whatever be ρ, provided the conditions (c) be fulfilled. This agrees with the result already obtained.

Either of the functions ϕ and ϕ', thus defined together, is called the *Conjugate* of the other : and when they are equal (i.e. when (c) is satisfied) ϕ is called a *Self-Conjugate* function. As we employed it in Chap. VI, ϕ was self-conjugate ; and, even had it not been so, it was involved (as we shall presently see) in such a manner that its non-conjugate part was necessarily absent.

We may now write, as before,

$$\phi\rho = -(a'Sa\rho + \beta'S\beta\rho + \gamma'S\gamma\rho),$$

and, by (d),

$$\phi'\rho = -(aSa'\rho + \beta S\beta'\rho + \gamma S\gamma'\rho).$$

From these we have by subtraction,

$$(\phi - \phi')\rho = \phi\rho - \phi'\rho = aSa'\rho - a'Sa\rho + \beta S\beta'\rho - \beta'S\beta\rho + \gamma S\gamma'\rho - \gamma'S\gamma\rho$$
$$= V.Vaa'\rho + V.V\beta\beta'\rho + V.V\gamma\gamma'\rho$$
$$= 2V.\,\epsilon\rho \dots\dots\dots\dots\dots\dots\dots\dots\dots\dots\dots\dots\dots\dots(g)\,;$$

if we agree to write

$$2\epsilon = V(aa' + \beta\beta' + \gamma\gamma') \dots\dots\dots\dots\dots\dots(h).$$

We may now express that ϕ is self-conjugate by writing

$$\epsilon = 0,$$

the physical interpretation of which equation is of the highest importance, as will soon appear.

If we form by means of (a) the value of ϵ as in (h) we get

$$2\epsilon = (c\gamma - b'\beta) + (aa - c'\gamma) + (b\beta - a'a)$$
$$= (a - a')\,a + (b - b')\,\beta + (c - c')\,\gamma,$$

which obviously cannot vanish unless (as before) the three conditions (c) are satisfied.

By adding the values of $\phi\rho$ and $\phi'\rho$ above we obtain

$$(\phi + \phi')\,\rho = \phi\rho + \phi'\rho = -\,(aSa'\rho + a'Sa\rho + \beta S\beta'\rho + \beta'S\beta\rho + \gamma S\gamma'\rho + \gamma'S\gamma\rho)$$

$$= -\,V\,(a\rho a' + \beta\rho\beta' + \gamma\rho\gamma') - \rho\,(Saa' + S\beta\beta' + S\gamma\gamma').$$

As we have (by 69. 6)

$$V\,.\,a\rho a' = V\,.\,a'\rho a,\ \ \&c.$$

this new function of ρ is self-conjugate.

This will easily be seen by putting $\phi + \phi'$ for ϕ in (b) and re-membering that (by 69. 17) we have

$$S\,.\,\sigma a\rho a' = S\,.\,\rho a'\sigma a = S\,.\,\rho a\sigma a',\ \&c.,\ \&c.$$

Hence we may write

$$(\phi + \phi')\,\rho = 2\bar{\varpi}\rho\dots\dots\dots\dots\dots\dots\dots(i),$$

where the bar over ϖ signifies that it is self-conjugate, and the factor 2 is introduced for convenience.

From (g) and (i) we have

$$\left.\begin{array}{l} \phi\rho = \bar{\varpi}\rho + V\epsilon\rho \\ \phi'\rho = \bar{\varpi}\rho - V\epsilon\rho \end{array}\right\}\dots\dots\dots\dots\dots\dots\dots(j).$$

If instead of $\phi\rho$ in any of the above investigations we write $(\phi + g)\,\rho$, it is obvious that $\phi'\rho$ becomes $(\phi' + g)\,\rho$: and the only change in the coefficients in (a) and (f) is the addition of g to each of the main series A, B, C.

We now come to Hamilton's grand proposition with regard to linear and vector functions. If ϕ be such that, in general, the vectors

$$\rho,\ \ \phi\rho,\ \ \phi^2\rho$$

(where $\phi^2\rho$ is an abbreviation for $\phi\,(\phi\rho)$) are not in one plane, then any fourth vector such as $\phi^3\rho$ (a contraction for $\phi\,(\phi(\phi\rho))$) can be expressed in terms of them as in 31. 5.

Thus $\qquad\qquad \phi^3\rho = m_2\phi^2\rho - m_1\phi\rho + m\rho\dots\dots\dots\dots\dots(k),$

where m, m_1, m_2 are scalars whose values will be found immedi-ately. That they are independent of ρ is obvious, for we may put

a, β, γ in succession for ρ and thus obtain three equations of the form

$$\phi^3 a = m_2 \phi^2 a - m_1 \phi a + m a \dots \dots \dots \dots (l),$$

from which their values can be found. For by repeated applications of (a) we can express (l) in the form

$$\overset{\vee\vee}{A} a + \overset{\vee\vee}{B} \beta + \overset{\vee\vee}{C} \gamma = 0.$$

This gives $\qquad \overset{\vee\vee}{A} = 0, \quad \overset{\vee\vee}{B} = 0, \quad \overset{\vee\vee}{C} = 0.$

These are three equations connecting m, m_1, m_2, with the nine coefficients in (a). The other two groups of three equations, furnished by the other two equations of the form (l), are merely *consistent* with these; and involve no farther limitations. This method, however, is very inferior to one which will shortly be given.

Conversely, if quantities m, m_1, m_2 can be found which satisfy (l), we may reproduce (k) by putting

$$\rho = x a + y \beta + z \gamma$$

and adding together the three expressions (l) multiplied by x, y, z respectively. For it is obvious from the expression for ϕ that

$$x \phi \rho = \phi (x \rho), \qquad x \phi^2 \rho = \phi^2 (x \rho), \ \&c.,$$

whatever scalar be represented by x.

If ρ, $\phi \rho$, and $\phi^2 \rho$ are in the same plane, then applying the strain ϕ again we find $\phi \rho$, $\phi^2 \rho$, $\phi^3 \rho$ in one plane; and thus equation (k) holds for this case also. And it of course holds if $\phi \rho$ is parallel to ρ, for then $\phi^2 \rho$ and $\phi^3 \rho$ are also parallel to ρ.

We will prove that scalars can be found which satisfy the three equations (l) (equivalent to *nine* scalar equations, of which, however, as we have seen, six depend upon the other three) by actually determining their values.

The volume of the parallelepiped whose three conterminous edges are λ, μ, ν is (31. 1)

$$- S . \lambda \mu \nu.$$

After the strain its volume is

$$- S . \phi\lambda \, \phi\mu \, \phi\nu,$$

so that the ratio

$$\frac{S . \phi\lambda \phi\mu \phi\nu}{S . \lambda\mu\nu}$$

is the same whatever vectors λ, μ, ν may be; and depends therefore on the constants of ϕ alone. We may therefore assume

$$\left. \begin{aligned} \lambda &= \rho, \\ \mu &= \phi\rho, \\ \nu &= \phi^2\rho, \end{aligned} \right\}$$

and by inspection of (k) we find

$$\frac{S . \phi\lambda \phi\mu \phi\nu}{S . \lambda\mu\nu} = \frac{S . \phi\rho\phi^2\rho\phi^3\rho}{S . \rho\phi\rho\phi^2\rho} = m \ldots\ldots\ldots\ldots (m),$$

which gives the physical meaning of this constant in (k). As we may put if we please

$$\left. \begin{aligned} \lambda &= a, \\ \mu &= \beta, \\ \nu &= \gamma, \end{aligned} \right\}$$

we see by (a) that

$$m = \frac{S . \phi a\phi\beta\phi\gamma}{S . a\beta\gamma} = \begin{vmatrix} A, & c, & b' \\ c', & B, & a \\ b, & a', & C \end{vmatrix},$$

which is the expression for the ratio in which the volume of each portion has been increased. This is unchanged by putting ϕ' for ϕ, for it becomes, by (f),

$$m = \begin{vmatrix} A, & c', & b \\ c, & B, & a' \\ b', & a, & C \end{vmatrix}.$$

Hence *conjugate strains produce equal changes of volume.*

Recurring to (m) we may write it by (e) as

$$S . \lambda\phi' V\phi\mu\phi\nu = mS . \lambda V\mu\nu,$$

from which, as λ is *absolutely any vector*, we have

$$\phi' V\phi\mu\phi\nu = m V\mu\nu$$

or
$$\phi V\phi'\mu\phi'\nu = m V\mu\nu \Big\} \dots\dots\dots\dots\dots(n).$$

[In passing we may notice that (n) gives us the complete solution of a linear and vector equation such as

$$\phi\sigma = \delta,$$

where δ and ϕ are given and σ is to be found. We have in fact only to take any two vectors μ and ν which are perpendicular to δ, and such that

$$V\mu\nu = \delta,$$

and we have for the unknown vector

$$\sigma = \frac{1}{m} V\phi'\mu\phi'\nu,$$

which can be calculated, as ϕ is given.]

If in (n) we put $\phi + g$ for ϕ we must do so for the value of m in (m). Calling the latter M_g we have

$$M_g = \frac{S . (\phi + g) \lambda (\phi + g) \mu (\phi + g) \nu}{S . \lambda\mu\nu}$$

$$= m + g \frac{S . \lambda\phi\mu\phi\nu + S . \mu\phi\nu\phi\lambda + S . \nu\phi\lambda\phi\mu}{S . \lambda\mu\nu}$$

$$+ g^2 \frac{S . \lambda\mu\phi\nu + S . \nu\lambda\phi\mu + S . \mu\nu\phi\lambda}{S . \lambda\mu\nu}$$

$$+ g^3 \dots\dots\dots\dots\dots\dots\dots\dots\dots\dots\dots\dots(o),$$

and by (n) $(\phi + g) V (\phi' + g) \mu (\phi' + g) \nu = M_g . V\mu\nu \dots\dots(p),$

or $M_g = m + \mu_1 g + \mu_2 g^2 + g^3$

$$(\phi + g)[m\phi^{-1} V\mu\nu + g (V\phi'\mu\nu + V\mu\phi'\nu) + g^2 V\mu\nu] = M_g V\mu\nu \Big\} ..(q).$$

From the latter of these equations it is obvious that

$$V\phi'\mu\nu + V\mu\phi'\nu$$

must be a linear and vector function of $V\mu\nu$, since all the other terms of the equation are such functions.

As practice in the use of these functions we will solve a problem of a little greater generality. The vectors

$$V\mu\nu, \quad V\phi'\mu\nu, \quad \text{and} \quad V\mu\phi'\nu$$

are not generally coplanar. In terms of these (31. 5), let us express $\phi V\mu\nu$.

Let $\qquad \phi V\mu\nu = xV\mu\nu + yV\phi'\mu\nu + zV\mu\phi'\nu.$

Operate by $S . \lambda$, $S . \mu$, $S . \nu$ successively, then

$$S . \mu\nu\phi'\lambda = xS . \lambda\mu\nu + yS . \nu\lambda\phi'\mu + zS . \lambda\mu\phi'\nu,$$
$$S . \mu\nu\phi'\mu = yS . \nu\mu\phi'\mu,$$
$$S . \mu\nu\phi'\nu = zS . \nu\mu\phi'\nu.$$

The two last equations give (by 69. 4)

$$y = -1, \quad z = -1,$$

and therefore the first gives

$$x = \frac{S . \mu\nu\phi'\lambda + S . \nu\lambda\phi'\mu + S . \lambda\mu\phi'\nu}{S . \lambda\mu\nu}$$
$$= \mu_2, \text{ by } (q).$$

Hence, finally,

$$\phi V\mu\nu = \mu_2 V\mu\nu - V\phi'\mu\nu - V\mu\phi'\nu \dots \dots \dots \dots (r).$$

Substituting this in (q), and putting σ for $V\mu\nu$, which is any vector whatever, we have

$$(\phi + g)\left[m\phi^{-1} + g(\mu_2 - \phi) + g^2\right]\sigma = (m + \mu_1 g + \mu_2 g^2 + g^3)\sigma,$$

or, multiplying out,

$$(m - g\phi^2 + \mu_2 g\phi - g^2\phi + gm\phi^{-1} + g^2\phi + g^2\mu_2 + g^3)\sigma$$
$$= (m + \mu_1 g + \mu_2 g^2 + g^3)\sigma;$$

that is $\qquad (-\phi^2 + \mu_2\phi + m\phi^{-1})\sigma = \mu_1\sigma,$

or $\qquad (\phi^3 - \mu_2\phi^2 + \mu_1\phi - m)\sigma = 0.$

Comparing this with (k) we see that

$$\left. \begin{array}{l} m_2 = \mu_2 = \dfrac{S . \lambda\mu\phi\nu + S . \nu\lambda\phi\mu + S . \mu\nu\phi\lambda}{S . \lambda\mu\nu} \\[3mm] m_1 = \mu_1 = \dfrac{S . \lambda\phi\mu\phi\nu + S . \mu\phi\nu\phi\lambda + S . \nu\phi\lambda\phi\mu}{S . \lambda\mu\nu} \end{array} \right\} \quad \dots\dots\dots(s),$$

and thus the determination is complete.

We may write (k), if we please, in the form

$$m\phi^{-1}\rho = m_1\rho - m_2\phi\rho + \phi^2\rho \quad \dots\dots\dots\dots(k'),$$

which gives another, and more direct, solution of the equation (above mentioned)

$$\phi\sigma = \delta.$$

Physically, the result we have arrived at is the solution of the problem, "By adding together scalar multiples of any vector of a body, of the corresponding vector of the same strained homogeneously, and of that of the same twice over strained, to represent the state of the body which would be produced by supposing the strain to be reversed or inverted."

These properties of the function ϕ are sufficient for many applications, of which we proceed to give a few.

I.   Homogeneous strain converts an originally spherical portion of a body into an ellipsoid.

For if ρ be a radius of the sphere, σ the vector into which it is changed by the strain, we have

$$\sigma = \phi\rho,$$

and $$T\rho = C,$$

from which we obtain

$$T\phi^{-1}\sigma = C,$$

or $$S \cdot \phi^{-1}\sigma\phi^{-1}\sigma = -C^2,$$

or, finally, $$S \cdot \sigma\phi'^{-1}\phi^{-1}\sigma = -C^2.$$

This is the equation of a central surface of the second degree; and, therefore, of course, from the nature of the problem, an ellipsoid.

II.   To find the vectors whose direction is unchanged by the strain.

Here $\phi\rho$ must be parallel to ρ or

$$\phi\rho = g\rho.$$

This gives $$\phi^2\rho = g^2\rho, \&c.,$$

so that by (k) we have

$$g^3 - m_2 g^2 + m_1 g - m = 0.$$

This must have one real root, and may have three. Suppose g_1 to be a root, then

$$\phi\rho - g_1\rho = 0,$$

and therefore, whatever be λ,

$$S\lambda\phi\rho - g_1 S\lambda\rho = 0,$$

or
$$S \cdot \rho\,(\phi'\lambda - g_1\lambda) = 0.$$

Thus it appears that the operator $\phi' - g_1$ cuts off from any vector λ the part which is parallel to the required value of ρ, and therefore that we have

$$\rho \parallel MV \cdot (\phi' - g_1)\,\lambda\,(\phi' - g_1)\,\mu$$
$$\parallel \{m\phi^{-1} - g_1\,(m_2 - \phi) + g_1{}^2\}\,\zeta,$$

where ζ is absolutely any vector whatever. This may be written as

$$\rho \parallel \left\{\frac{m}{g_1} - (m_2 - g_1)\,\phi + \phi^2\right\}\zeta$$
$$\parallel \frac{\phi^3 - m_2\phi^2 + m_1\phi - m}{\phi - g_1}\,\zeta.$$

The same result may more easily be obtained thus :—

The expression

$$\left(\phi^3 - m_2\phi^2 + m_1\phi - m\right)\rho = 0,$$

being true for all vectors whatever, may be written

$$(\phi - g_1)\,(\phi - g_2)\,(\phi - g_3)\,\rho = 0,$$

and it is obvious that each of these factors deprives ρ of the portion corresponding to it : i.e. $\phi - g_1$ applied to ρ cuts off the part parallel to the root of

$$(\phi - g_1)\,\sigma = 0, \text{ &c., &c.}$$

so that the operator $(\phi - g_2)\,(\phi - g_3)$ when applied to a vector leaves only that part of it which is parallel to σ where

$$(\phi - g_1)\,\sigma = 0.$$

III. Thus it appears that there is always one vector, and that there may be three vectors, whose direction is unchanged by the strain.

DEF. *Pure, or non-rotational, strain consists in altering the lengths of three lines at right angles to one another, without altering their directions.*

Hence if
$$\phi\rho_1 = g_1\rho_1$$
$$\phi\rho_2 = g_2\rho_2$$
$$\phi\rho_3 = g_3\rho_3$$

the strain ϕ is pure if, and not unless, ρ_1, ρ_2, ρ_3 form a rectangular system. [There is a qualification if two or more of $g_1 g_2 g_3$ be equal.]

Hence, for a pure strain, we have
$$S\rho_2\phi\rho_3 = g_1 S\rho_2\rho_1 = 0,$$
and
$$S\rho_1\phi\rho_2 = g_2 S\rho_1\rho_2 = 0 :$$
or
$$S\rho_1\phi\rho_2 = S\rho_2\phi\rho_1.$$

But we have, generally,
$$S\rho_1\phi\rho_2 = S\rho_2\phi'\rho_1.$$

As we have two other pairs of equations like these, we see that
$$\phi = \phi'$$
when the strain is pure.

Conversely, if
$$\phi = \phi'$$
the three unchanging directions ρ_1, ρ_2, ρ_3 are perpendicular to one another.

For, in this case, the roots of
$$M_g = 0$$
are real. Let them be such that
$$\left.\begin{array}{l} (\phi - g_1)\,\rho_1 = 0 \\ (\phi - g_2)\,\rho_2 = 0 \\ (\phi - g_3)\,\rho_3 = 0 \end{array}\right\},$$

then $g_1 g_2 S\rho_1\rho_2 = S\phi\rho_1\phi\rho_2$

$$= S\rho_1 \phi\phi\rho_2$$

(because, by hypothesis, the strain is pure)

$$= g_2{}^2 S\rho_1\rho_2,$$

for $\phi\rho_2 = g_2\rho_2$ and $\phi^2\rho_2 = g_2{}^2\rho_2$.

Hence, except in the particular case of

$$g_1 = g_2,$$

we must have

$$S\rho_1\rho_2 = 0,$$

whence the proposition.

When g_1 and g_2 are equal, ρ_1 and ρ_2 are each perpendicular to ρ_3, but *any* vector in their plane satisfies

$$\phi\sigma - g_1\sigma = 0.$$

When all three roots are equal, *every* vector satisfies

$$\phi\sigma - g_1\sigma = 0.$$

IV.   Thus we see that when the strain is unaccompanied by rotation the three values of g are real. [But we must take care to notice that the converse does not hold. This will be discussed later.] If these values be real and *different*, there are three vectors at right angles to one another which are the only lines in the body whose directions remain unchanged. When two are equal, every vector parallel to a given plane, and all vectors perpendicular to it, are unchanged in direction. When all three are equal no vector has its direction changed.

V.   There is, however, a peculiarity to be noticed, which distinguishes true physical strain from the results of our mathematical analysis. When one or more of the values of g has a *negative* sign, we cannot interpret *physically* the result without introducing the idea of a pure strain which shall, as it were, pull the parts of an originally spherical portion of the body through the centre of the sphere, and so form an ellipsoid by turning a part of the body outside in. When two, only, are negative we can represent physically the result by introducing the conception of a rotation through two right angles about the third axis. But we began by assuming that there is no rotation! Hence, for the case considered, all three roots must be positive. See end of next section (VI.).

VI. This will appear more clearly if we take the case of a rigid body, for here we must have, whatever vectors be represented by ρ and σ,

$$\left.\begin{array}{l} T\phi\rho = T\rho \\ S\rho\sigma = S . \phi\rho\phi\sigma \end{array}\right\} \dots\dots\dots\dots\dots\dots\dots (t),$$

i. e. the lengths of vectors, and their inclinations to one another, are unaltered. In this case, therefore, the strain can be nothing but a rotation. It is easy to see that the second of these equations includes the first; so that if, for variety, we take ϕ as represented in equations (a), and write

$$\rho = x\alpha + y\beta + z\gamma,$$
$$\sigma = \xi\alpha + \eta\beta + \zeta\gamma,$$

we have, for *all* values of the six scalars x, y, z, ξ, η, ζ, the following identity :

$$- (x\xi + y\eta + z\zeta) = S . (x\alpha' + y\beta' + z\gamma') (\xi\alpha' + \eta\beta' + \zeta\gamma')$$
$$= \alpha'^2 x\xi + \beta'^2 y\eta + \gamma'^2 z\zeta$$
$$+ (x\eta + y\xi) S\alpha'\beta' + (y\zeta + z\eta) S\beta'\gamma' + (z\xi + x\zeta) S\gamma'\alpha'.$$

This necessitates

$$\left.\begin{array}{l} \alpha'^2 = \beta'^2 = \gamma'^2 = -1 \\ S\alpha'\beta' = S\beta'\gamma' = S\gamma'\alpha' = 0 \end{array}\right\} \dots\dots\dots\dots\dots (u),$$

i. e. the vectors α', β', γ' form, like α, β, γ, a rectangular unit system. And it is evident that *any* and *every* such system satisfies the given conditions. But the system α', β', γ' must be similar to α, β, γ, i.e. if a quadrant of *positive* rotation round α changes β to γ &c. a quadrant of *positive* rotation about α' must change β' to γ' &c.

When this is not the case, the system α', β', γ' is the *per-*

version of a, β, γ, i.e. its image in a plane mirror; and the strain is impossible from a physical point of view.

This is easily seen from another point of view. The volume of the parallelepiped whose edges are rectangular unit vectors a, β, γ is
$$- S . a\beta\gamma$$
if a positive quadrant of rotation round a brings β to coincide with γ &c. But, in the perverted system, *the volume has changed sign* and is expressed by
$$S . a\beta\gamma.$$

VII. It may be interesting to form, for this particular case, the equation giving the values of g. We have
$$M_g = \frac{S . (\phi + g)\, a\, (\phi + g)\, \beta\, (\phi + g)\, \gamma}{S . a\beta\gamma}$$
$$= \frac{S . (a' + ga)\, (\beta' + g\beta)\, (\gamma' + g\gamma)}{S . a\beta\gamma}$$
$$= 1 - gS\, (a\beta'\gamma' + a'\beta\gamma' + a'\beta'\gamma)$$
$$\qquad - g^2 S\, (a\beta\gamma' + a\beta'\gamma + a'\beta\gamma) + g^3.$$

Recollecting that a, β, γ; a', β', γ' are systems of rectangular unit vectors, we find that this may be written
$$M_g = 1 - (g + g^2)\, S\, (aa' + \beta\beta' + \gamma\gamma') + g^3$$
$$= (g + 1)\, [g^2 - g\, \{1 + S\, (aa' + \beta\beta' + \gamma\gamma')\} + 1].$$

Hence the roots of
$$M_g = 0$$
are in this case; first and always,
$$g_1 = -1,$$
which refers to the axis about which the rotation takes place: secondly, the roots of
$$g^2 - g\, \{1 + S\, (aa' + \beta\beta' + \gamma\gamma')\} + 1 = 0.$$

Now the roots of this equation are imaginary so long as the coefficient of the first power of g lies *between* the limits ± 2.

Also the values of the several quantities Saa', $S\beta\beta'$, $S\gamma\gamma'$ can never exceed the limits ± 1. When the system a, β, γ coincides with α', β', γ', the value of each of the scalars is -1, and the coefficient of the first power of g is $+2$. When two of them are equal to $+1$ and the third to -1 we have the coefficient of the first power of $g = -2$. These are the only two cases in which the three values of g are all real.

In the first, all three values of g are equal to -1, i.e.

$$\phi\rho = \rho$$

for all values of ρ, and there is no rotation whatever. In the second case there is a rotation through two right angles about the axis of the -1 value of g.

VIII. It is an exceedingly remarkable fact that, however a body may be homogeneously strained, there is always at least one vector whose direction remains unchanged. The proof is simply based on the fact that the strain-function depends on a cubic equation (with real coefficients) which must have at least one real root.

IX. As an illustration of what precedes (though one which must be approached cautiously), suppose a body to be strained so that three vectors, α'', β'', γ'' (not coplanar, and not necessarily at right angles to one another), preserve their direction, becoming $e_1\alpha''$, $e_2\beta''$, $e_3\gamma''$. Then we have

$$\phi\rho S . \alpha''\beta''\gamma'' = e_1\alpha''S . \beta''\gamma''\rho + e_2\beta''S . \gamma''\alpha''\rho + e_3\gamma''S . \alpha''\beta''\rho.$$

By the formulæ (m, s) we have

$$m = \frac{S . \phi\alpha''\phi\beta''\phi\gamma''}{S . \alpha''\beta''\gamma''} = e_1e_2e_3,$$

$$m_1 = \frac{S\left(\alpha''\phi\beta''\phi\gamma'' + \beta''\phi\gamma''\phi\alpha'' + \gamma''\phi\alpha''\phi\beta''\right)}{S . \alpha''\beta''\gamma''} = e_2e_3 + e_3e_1 + e_1e_2,$$

$$m_2 = \frac{S\left(\alpha''\beta''\phi\gamma'' + \beta''\gamma''\phi\alpha'' + \gamma''\alpha''\phi\beta''\right)}{S . \alpha''\beta''\gamma''} = e_1 + e_2 + e_3;$$

so that we have by (k)

$$(\phi - e_1)(\phi - e_2)(\phi - e_3)\rho = 0.$$

Though the values of g are here all real, we must not rashly adopt the conclusions of (IV.), for we must remember that α'', β'', γ'' do not, like α, β, γ, necessarily form a rectangular system.

In this case we have

$$\phi'\rho S \cdot \alpha''\beta''\gamma'' = e_1 V\beta''\gamma'' S\alpha''\rho + e_2 V\gamma''\alpha'' S\beta''\rho + e_3 V\alpha''\beta'' S\gamma''\rho.$$

So that, by (h),

$$2\epsilon S \cdot \alpha''\beta''\gamma'' = V \cdot (e_1\alpha'' V\beta''\gamma'' + e_2\beta'' V\gamma''\alpha'' + e_3\gamma'' V\alpha''\beta'')$$
$$= (\overline{e_2 - e_3}\alpha'' S\beta''\gamma'' + \overline{e_3 - e_1}\beta'' S\gamma''\alpha'' + \overline{e_1 - e_2}\gamma'' S\alpha''\beta'').$$

This vanishes, or the strain is pure, if either

1. $$S\alpha''\beta'' = S\beta''\gamma'' = S\gamma''\alpha'' = 0,$$

i.e. if α'', β'', γ'' are rectangular, in which case e_1, e_2, e_3 may have any values; or

2. $e_1 = e_2 = e_3$, in which case

$$\phi'\rho S \cdot \alpha''\beta''\gamma'' = e_1\{V\beta''\gamma'' S\alpha''\rho + V\gamma''\alpha'' S\beta''\rho + V\alpha''\beta'' S\gamma''\rho\}$$
$$= e_1\rho S \cdot \alpha''\beta''\gamma'' \text{ by (69. 14),}$$

so that

$$\phi'\rho = e_1\rho = \phi\rho$$

for every vector: a general uniform dilatation unaccompanied by change of direction.

3. $e_1 = e_2$, and α'' and β'' both perpendicular to γ''.

From what precedes it is evident that for the complete study of a strain we must endeavour to distinguish in each case between the *pure* strain and the merely *rotational* part. If a strain be capable of being decomposed into 1st a pure strain, 2nd a rotation, it is obvious that the vectors which in the altered state of the body become the axes of the strain-ellipsoid (I.) must have been originally at right angles to one another.

The equation of the strain-ellipsoid is

$$S\rho\phi^{-2}\rho = -c^2,$$

and in this it is obvious that ϕ^{-2} is self-conjugate, or at least is to be treated as such: for a non-conjugate term in $\phi^{-2}\rho$ would be (g) of the form $$V\epsilon\rho,$$

and would therefore not appear in the equation.

For the proper treatment of rotations, the following simple but excessively important proposition, due to Hamilton, forms the best starting-point.

If q be any quaternion, the operator q ( ) q^{-1} turns the vector, quaternion, or body operated on round an axis perpendicular to the plane of q and through an angle equal to double that of q.

For the proof we refer the reader to Hamilton's *Lectures*, § 282, *Elements*, § 179 (1), or Tait, § 353. It is obvious that the tensor of q may be taken to be unity, i.e. q may be considered as a mere versor, because the value of its tensor does not affect that of the operator.

[A very simple but important example of this proposition is given by supposing q and r to be both vectors, a and β let us say. Then

$$a\beta a^{-1}$$

is the result of turning β conically through two right angles about a, i.e. if a be the normal to a reflecting surface and β the incident ray, $-a\beta a^{-1}$ is the reflected ray.]

Now let the strain ϕ be effected by (1), a pure strain ϖ (self-conjugate of course) followed by the rotation q ( ) q^{-1}. We have, for all values of ρ,

$$\phi\rho = q\,(\varpi\rho)\,q^{-1}\dots\dots\dots\dots\dots\dots\dots(v).$$

whence $$\phi'\rho = \varpi\,(q^{-1}\rho q).$$

The interpretation is that, under the above definition, *the conjugate to any strain consists of the reversed rotation, followed by the pure strain.*

We may of course put, as in Chap. VI,

$$\varpi\rho = e_1 aS a\rho + e_2 \beta S \beta\rho + e_3 \gamma S \gamma\rho,$$

where a, β, γ form a rectangular system. Hence

$$\phi\rho = e_1 q a q^{-1} S a\rho + e_2 q\beta q^{-1} S\beta\rho + e_3 q\gamma q^{-1} S\gamma\rho.$$

Here the axes are parallel to

$$qaq^{-1}, \quad q\beta q^{-1}, \quad q\gamma q^{-1},$$

and we have

$$S \cdot qaq^{-1}q\beta q^{-1} = S \cdot qa\beta q^{-1} = Sa\beta = 0, \&c.$$

So far the matter is nearly self-evident, but we now come to the important question of the *separation of the pure strain from the rotation.* By the formulæ above we see that

$$\phi'\phi\rho = \overline{\omega}q^{-1}\phi\rho q$$
$$= \overline{\omega}q^{-1}(q\overline{\omega}\rho q^{-1})q$$
$$= \overline{\omega}^2\rho,$$

so that we have in symbols, for the determination of $\overline{\omega}$, the equation

$$\phi'\phi = \overline{\omega}^2.$$

That is, as we see at once from the statements above, *any strain, followed by its conjugate, gives a pure strain, which is the square (or the result of two applications) of the pure part of either.*

To solve this equation we employ expressions like (k). $\phi'\phi$ being a known function, let us call it ω, and form its equation as

$$\omega^3 - m_2\omega^2 + m_1\omega - m = 0.$$

Here the coefficients are perfectly determinate.

Also *suppose* that the corresponding equation in $\overline{\omega}$ is

$$\overline{\omega}^3 - g_2\overline{\omega}^2 + g_1\overline{\omega} - g = 0,$$

where g, g_1, g_2 are unknown scalars. By the help of the given relation $\qquad\qquad \overline{\omega}^2 = \omega,$

we may modify this last equation as follows:

$$\overline{\omega}\omega - g_2\omega + g_1\overline{\omega} - g = 0,$$

whence $\qquad\qquad \overline{\omega} = \dfrac{g + g_2\omega}{g_1 + \omega};$

i.e. $\overline{\varpi}$ is given definitely in terms of the known function ω, as soon as the quantities g are found. But our given equation

$$\overline{\varpi}^2 = \omega$$

may now be written

$$\left(\frac{g + g_2\omega}{g_1 + \omega}\right)^2 = \omega,$$

or $\qquad \omega^3 - \left(g_2^2 - 2g_1\right)\omega^2 + \left(g_1^2 - 2gg_2\right)\omega - g^2 = 0.$

As this is an equation between ω and constants it must be equivalent to that already given: so that, comparing coefficients, we have

$$g_2^2 - 2g_1 \ = m_2,$$
$$g_1^2 - 2gg_2 = m_1,$$
$$g^2 \qquad\quad = m \ ;$$

from which, by elimination of g and g_2, we have

$$\left(\frac{g_1^2 - m_1}{2\sqrt{m}}\right)^2 = m_2 + 2g_1 \ .$$

The solution of the problem is therefore reduced to that of this biquadratic equation; for, when g_1 is found, g_2 is given linearly in terms of it.

It is to be observed that in the operations above we have not been particular as to the arrangement of factors. This is due to the fact that any functions of the *same* operator are commutative in their application.

Having thus found the pure part of the strain we have at once the rotation, for (v) gives

$$\phi\overline{\varpi}^{-1} \rho = q\rho q^{-1},$$

or, as it may more expressively be written,

$$\frac{\phi}{\sqrt{\phi'\phi}} = q \left(\ \ \right) q^{-1}.$$

If instead of (v) we write

$$\phi\rho = \overline{\omega}\left(r\rho r^{-1}\right)\ldots\ldots\ldots\ldots\ldots\ldots\ldots(v'),$$

we assume that the rotation takes place first, and is succeeded by the pure strain. This form gives

$$\phi'\rho = r^{-1}\,(\bar{\omega}\rho)\,r,$$

and

$$\phi\phi'\rho = \bar{\omega}^2\rho,$$

whence $\bar{\omega}$ is found as above. And then (v') gives

$$\bar{\omega}^{-1}\phi = r\,(\quad)\,r^{-1}.$$

Thus, to recapitulate, a strain ϕ is equivalent to the pure strain $\sqrt{\phi'\phi}$ followed by the rotational strain $\phi\,\dfrac{1}{\sqrt{\phi'\phi}}$, or to the rotational strain $\dfrac{1}{\sqrt{\phi\phi'}}\,\phi$ followed by the pure strain $\sqrt{\phi\phi'}$.

This leads us, as an example, *to find the condition that a given strain is rotational only*, i.e. that a quaternion q can be found such that

$$\phi = q\,(\quad)\,q^{-1}.$$

Here we have

$$\phi' = q^{-1}\,(\quad)\,q,$$

or

$$\phi' = \phi^{-1} \dotfill (w).$$

But

$$m\phi^{-1} = m_1 - m_2\phi + \phi^2,$$

or

$$m\phi' = m_1 - m_2\phi + \phi^2,$$
$$\text{whose conjugate is} \quad m\phi = m_1 - m_2\phi' + \phi'^2,$$

and the elimination of ϕ' between these two equations gives

$$m\phi = m_1 - \frac{m_2}{m}\,(m_1 - m_2\phi + \phi^2) + \frac{1}{m^2}\,(m_1 - m_2\phi + \phi^2)^2,$$

i.e.

$$0 = \begin{vmatrix} (m^2m_1 - mm_1m_2 + m_1{}^2) \\ -(m^3 - mm_2{}^2 + 2m_1m_2)\,\phi \\ -(mm_2 - 2m_1 - m_2{}^2)\,\phi^2 \\ -2m_2\phi^3 \\ +\phi^4 \end{vmatrix} = \begin{vmatrix} (m^2m_1 - mm_1m_2 + m_1{}^2) \\ -(m^3 - mm_2{}^2 + 2m_1m_2 - m)\,\phi \\ +(2m_1 + m_2{}^2 - mm_2 - m_1)\,\phi^2 \\ -m_2\phi^3 \end{vmatrix}$$

by using the expression for ϕ^4 from the cubic in ϕ.

Now this last expression can be nothing else than the cubic in ϕ itself, else ϕ would have *two different* sets of constants in the form (k), which is absurd, as these constants, from the mode in which they are determined, can have but single values. Thus we have, by comparing coefficients,

$$\left. \begin{array}{l} m_2^{\,2} = 2m_1 + m_2^{\,2} - mm_2 - m_1 \\ m_1 m_2 = m^3 - mm_2^{\,2} + 2m_1 m_2 - m \\ mm_2 = m^2 m_1 - mm_1 m_2 + m_1^{\,2} \end{array} \right\} .$$

The first gives

$$m_1 = mm_2,$$

by the help of which the second and third each become

$$m^3 - m = 0.$$

The value

$$m = 0$$

is to be rejected, as otherwise we should have been working with non-existent terms; and m, as the ratio of the volumes of two tetrahedra, is positive, so that finally

$$m = 1,$$

$$m_1 = m_2,$$

and the cubic for a rotational strain is, therefore,

$$\phi^3 - m_2 \phi^2 + m_2 \phi - 1 = 0,$$

or

$$(\phi - 1)\{\phi^2 + (1 - m_2)\,\phi + 1\} = 0,$$

where m_2 is left undetermined.

By comparison with the result of (VII.) we see that in the notation there employed

$$m_2 = - S\,(\alpha\alpha' + \beta\beta' + \gamma\gamma').$$

The student will perhaps here require to be reminded that in the section just referred to we employed the positive sign in operators such as $\phi + g$. In the one case the coefficients in the cubic are all positive, in the other they are alternately positive and negative. The example we have given is a particularly valuable one, as it gives a glimpse of the extent to which the separation of symbols can be safely carried in dealing with these questions.

DEF. A *simple shear* is a homogeneous strain in which all planes parallel to a fixed plane are displaced in the same direction parallel to that plane, and therefore through spaces proportional to their distances from that plane.

Let α be normal to the plane, β the direction of displacement, the former being considered as an unit-vector, and the tensor of the latter being the displacement of points at unit distance from the plane.

We obviously have, by the definition,

$$S\alpha\beta = 0.$$

Now if ρ be the vector of any point, drawn from an origin in the fixed plane, the distance of the point from the plane is

$$-S\alpha\rho.$$

Hence, if σ be the vector of the point after the shear,

$$\sigma = \phi\rho = \rho - \beta S\alpha\rho.$$

This gives

$$\phi'\rho = \rho - \alpha S\beta\rho,$$

which may be written as

$$= \rho - T\beta . \alpha S . U\beta\rho,$$

so that the conjugate of a simple shear is another simple shear equal to the former. But the direction of displacement in each shear is perpendicular to the unaltered planes in the other.

The equation for ϕ is easily found (by calculating m, m_1, m_2 from (m), (s)) to be

$$\phi^3 - 3\phi^2 + 3\phi - 1 = 0.$$

Putting $\phi'\phi = \psi$, we easily find (with $b = T\beta$)

$$\psi^3 - (3 + b^2)\psi^2 + (3 + b^2)\psi - 1 = 0.$$

Solving by the process lately described, we find

$$\left(\frac{g_1^2 - 3 - b^2}{2}\right)^2 = 3 + b^2 + 2g_1.$$

If $b = 2$, this gives $g_1 = 1$, and the farther equation

$$g_1{}^3 + g_1{}^2 - 13g_1 - 21 = 0,$$

of which $g_1 = -3$ is a root, so that

$$g_1{}^2 - 2g_1 - 7 = 0,$$

and $$g_1 = 1 \pm 2\sqrt{2}.$$

We leave to the student the selection (by trial) of the proper root, and the formation of the complete expressions for the pure and rotational parts of the strain in this simple and yet very interesting case.

As a simple example of the case in which two of the roots of the cubic are unreal, take the vector function when the strain is equivalent to a rotation θ about the unit vector a; the others of the rectangular system being β, γ.

Here we have, obviously,

$$\phi a = a,$$
$$\phi \beta = \beta \cos \theta + \gamma \sin \theta,$$
$$\phi \gamma = \gamma \cos \theta - \beta \sin \theta,$$

whence at once

$$- \phi\rho = aSa\rho + (\beta \cos \theta + \gamma \sin \theta)\,S\beta\rho + (\gamma \cos \theta - \beta \sin \theta)\,S\gamma\rho$$
$$= (1 - \cos \theta)\,aSa\rho - \rho \cos \theta - Va\rho \sin \theta.$$

Forming the quantities m, m_1, m_2 as usual, we have

$$\phi^3 - (1 + 2 \cos \theta)\,\phi^2 + (1 + 2 \cos \theta)\,\phi - 1 = 0,$$

or $$(\phi - 1)(\phi^2 - 2 \cos \theta\, \phi + 1) = 0,$$

or $$(\phi - 1)(\phi - \cos \theta - \sqrt{-1}\, \sin \theta)(\phi - \cos \theta + \sqrt{-1}\, \sin \theta) = 0.$$

Now

$$-(\phi - 1)\rho = (1 - \cos \theta)(aSa\rho + \rho) - \sin \theta\, Va\rho,$$

$$-(\phi - \cos \theta - \sqrt{-1}\, \sin \theta)\rho = (1 - \cos \theta)\,aSa\rho + \sin \theta\,(\rho\sqrt{-1} - Va\rho),$$

$$-(\phi - \cos \theta + \sqrt{-1}\, \sin \theta)\rho = (1 - \cos \theta)\,aSa\rho - \sin \theta\,(\rho\sqrt{-1} + Va\rho).$$

To detect the components which are destroyed by each of these factors separately, we have, by (II.), for $(\phi - 1)$, the vector

$$(\phi^2 - 2\cos\theta\,\phi + 1)\,\rho = -2aSa\rho\,(1 - \cos\theta)\,;$$

so that
$$(\phi - 1)\,a = 0,$$

which is, of course, true. Again

$$(\phi - 1)(\phi - \cos\theta - \sqrt{-1}\,\sin\theta)\rho = -\sin\theta\,(1 - \epsilon^{-\theta\sqrt{-1}})\,(\sqrt{-1}\,a + 1)\,Va\rho,$$

which we leave to the student to verify. The imaginary directions which correspond to the unreal roots are thus, in this case, parallel to the *Bivectors*

$$(a \pm \sqrt{-1})\,Va\rho.$$

Here, however, we reach notions which, though by no means difficult, cannot well be called elementary.

A very curious case, whose special interest however is rather mathematical than physical, is presented by the assumptions

$$a' = \beta + \gamma,$$
$$\beta' = \gamma + a,$$
$$\gamma' = a + \beta,$$

for then
$$\phi\rho = (\beta + \gamma)\,Sa\rho + (\gamma + a)\,S\beta\rho + (a + \beta)\,S\gamma\rho$$
$$= (a + \beta + \gamma)\,S(a + \beta + \gamma)\,\rho - (aSa\rho + \beta S\beta\rho + \gamma S\gamma\rho)$$
$$= 3\delta S\delta\rho + \rho,$$

where δ is a known unit vector. This function is obviously self-conjugate. Its cubic is

$$\phi^3 - 3\phi + 2 = 0 = (\phi - 1)^2(\phi + 2),$$

which might easily have been seen from the facts that

1st, $\phi\delta = -2\delta,$

2nd, $\phi a = a,$ if $Sa\delta = 0.$

The case is but slightly altered when the *signs* of a', β', γ' are changed. Then

$$\phi\rho = -3\delta S\delta\rho - \rho,$$

and the cubic is

$$\phi^3 - 3\phi - 2 = (\phi + 1)^2(\phi - 2) = 0.$$

These are mere particular cases of extension parallel to the single axis δ. The general expression for such extension is obviously

$$\phi\rho = \rho - e\delta S\delta\rho,$$

and we have for its cubic

$$(\phi - 1)^2 \{\phi - (1 + e)\} = 0.$$

We will conclude our treatment of strains by solving the following problem : *Find the conditions which must be satisfied by a simple shear which is capable of reducing a given strain to a pure strain.*

Let ϕ be the given strain, and let the shear be, as above,

$$\psi = 1 + \beta S \cdot a,$$

then the resultant strain is

$$\psi\phi = \phi + \beta S \cdot a\phi,$$
$$= \phi + \beta S \cdot \phi' a.$$

Taking the conjugate and subtracting, we must have

$$0 = \psi\phi - \phi'\psi' = \phi - \phi' + \beta S \cdot \phi'a - \phi'aS \cdot \beta$$
$$= 2V \cdot \epsilon - V \cdot (V\phi'a\,\beta),$$

so that the requisite conditions are contained in the sole equation

$$2\epsilon = V\phi'a\,\beta.$$

This gives   (1) $S \cdot \beta\epsilon = 0,$

(2) $S\phi'a\epsilon = 0 = Sa\phi\epsilon.$

But   (3) $Sa\beta = 0$ (by the conditions of a shear),

so that $xa = V \cdot \beta\phi\epsilon.$

Again, (4) $2\epsilon^2 = S \cdot \phi'a\,\beta\epsilon = S \cdot a\phi\,(\beta\epsilon)$
$$2x\epsilon^2 = S \cdot \beta\phi\epsilon\,\phi\,(\beta\epsilon) = -m\beta^2\epsilon^2,$$

or $-ma = 2V \cdot \beta^{-1}\phi\epsilon.$

Hence we may assume any vector perpendicular to ϵ for β, and a is immediately determined.

When two of the roots of the cubic in ϕ are imaginary let us suppose the three roots to be

$$e_1, \ e_2 \pm e_3 \sqrt{-1}.$$

Let β and γ be such that

$$\phi(\beta + \gamma\sqrt{-1}) = (e_2 + e_3\sqrt{-1})(\beta + \gamma\sqrt{-1}).$$

Then it is obvious that, by changing throughout the sign of the imaginary quantity, we have

$$\phi(\beta - \gamma\sqrt{-1}) = (e_2 - e_3\sqrt{-1})(\beta - \gamma\sqrt{-1}).$$

These two equations, when expanded, unite in giving by equating the real and imaginary parts the values

$$\left. \begin{array}{l} \phi\beta = e_2\beta - e_3\gamma \\ \phi\gamma = e_2\gamma + e_2\beta \end{array} \right\}.$$

To find the values of a, β, γ we must, as before, operate on any vector by two of the factors of the cubic.

As an example, take the very simple case

$$\phi\rho = e\,V i\rho.$$

Here it is easily seen by (m), (s), that $m = 0$, $m_1 = + e^2$, $m_2 = 0$, so that

$$\phi^3 + e^2\phi = 0,$$

that is

$$\phi(\phi + e\sqrt{-1})(\phi - e\sqrt{-1}) = 0.$$

As operand take

$$\rho = ix + jy + kz,$$

then

$$a \,\|\, V(\phi + e\sqrt{-1})(\phi - e\sqrt{-1})\,\rho$$
$$\|\, eV.(\phi + e\sqrt{-1})(ky - jz - \rho\sqrt{-1})$$
$$\|\, (-jy - kz + \rho)$$
$$\|\, i.$$

Again

$$\beta - \gamma\sqrt{-1} \,\|\, \phi(\phi + e\sqrt{-1})\,\rho$$
$$\|\, e\phi(ky - jz + \sqrt{-1}\rho)$$
$$\|\, -jy - kz + \sqrt{-1}(ky - jz)$$
$$\|\, jy + kz - \sqrt{-1}(jz - ky).$$

With a change of sign in the imaginary part, this will represent

$$\beta + \gamma \sqrt{-1},$$

so that
$$\beta = jy + kz,$$
$$\gamma = jz - ky.$$

Thus, as the student will easily find by trial, β and γ form with a a rectangular system. But for all that the system of principal vectors of ϕ, viz.

$$a, \; \beta \pm \gamma \sqrt{-1}$$

does not satisfy the conditions of rectangularity. In fact we see by the above values of β and γ that

$$S \cdot (\beta + \gamma \sqrt{-1}) (\beta - \gamma \sqrt{-1}) = \beta^2 + \gamma^2 = -2 \, (y^2 + z^2).$$

It may be well to call the student's attention at this point to the fact that the tensors of these imaginary vectors vanish, for

$$T^2 (\beta \pm \gamma \sqrt{-1}) = - S (\beta \pm \gamma \sqrt{-1}) (\beta \pm \gamma \sqrt{-1}) = \gamma^2 - \beta^2 = 0.$$

This gives a simple example of the new and very curious modifications which our results undergo when we pass to *Bivectors;* or, more generally, to *Biquaternions.*

As a pendant to the last problem we may investigate the relation of two vector-functions whose successive application produces rotation merely.

Here
$$\phi = \psi \chi^{-1}$$

is such that by (w)
$$\phi' = \phi^{-1},$$
$$\text{i.e. } \chi'^{-1} \psi' = \chi \psi^{-1},$$
or
$$\chi' \chi = \psi' \psi = \overline{\varpi}^3,$$

since each of these functions is evidently self-conjugate. This shews that the pure parts of the strains ψ and χ are the same, which is the sole condition.

One solution is, obviously,

$$\chi' = \chi^{-1}, \; \psi' = \psi^{-1},$$

T. Q.   14

i.e. each of the two is itself a rotation ; and a new proof that any number of successive rotations can be compounded into a single one may easily be given from this.

But we may also suppose either of ψ, χ, suppose the latter, to be self-conjugate, so that

$$\chi' = \chi = \bar{\chi},$$
or
$$\psi'\psi = \bar{\chi}^2,$$

which leads to previous results.

EXAMPLES TO CHAPTER X.

1. If α, β, γ be a rectangular unit system

$$S . V\alpha\phi\alpha V\beta\phi\beta V\gamma\phi\gamma = - mS . \beta\phi'^{-1}\alpha S . \beta (\phi - \phi') \alpha,$$

and therefore vanishes if ϕ be self-conjugate. State in words the theorem expressed by its vanishing.

2. With the same supposition find the values of

$$\Sigma V . V\alpha\phi\alpha . V\beta\phi\beta \text{ and of } \Sigma S . V\alpha\phi\alpha V\beta\phi\beta.$$

Also of $\qquad \Sigma . \alpha S\alpha\phi\alpha.$

3. When are two simple shears commutative ?

4. Expand $\dfrac{1}{1 - e\phi}$ in powers of ϕ, and reduce the result to three terms by the cubic in ϕ.

5. Shew that $\phi' V . \phi\rho\phi^2\rho = \dfrac{S . \phi\rho\phi^2\rho\phi^3\rho}{S . \rho\phi\rho\phi^2\rho} V . \rho\phi\rho$

$$= m V\rho\phi\rho.$$

6. Why cannot we expand ϕ' in terms of ϕ^0, ϕ, ϕ^2?

7. Express $V\rho\phi\rho$ in terms of ρ, $\phi\rho$, $\phi^2\rho$, and from the result find the conditions that $\phi\rho$ shall be parallel to ρ.

8. Given the coefficients of the cubic in ϕ, find those of the cubics in ϕ^2, ϕ^3, &c. ϕ^n.

9. Prove
$$\phi V . a\phi'a - mV . a\phi'^{-1}a = 0,$$
$$(\phi + m_2) V . a\phi'a = Va\phi'^2a.$$

10. If $m = \begin{vmatrix} A, & b, & c \\ a, & B, & c' \\ a', & b', & C \end{vmatrix}$ shew that $M_g = 0$ may be written as

$$\left\{ g^3 \frac{d^3}{dAdBdC} + g^2 \left(\frac{d^2}{dAdB} + \dots \right) + g \left(\frac{d}{dA} + \dots \right) + 1 \right\} m = 0,$$
$$\text{or } \epsilon^g \left(\frac{d}{dA} + \dots \right) m = 0.$$

11. Interpret the invariants m_1 and m_2 in connexion with Homogeneous Strain.

12. The cubics in $\phi\psi$ and $\psi\phi$ are the same.

13. Find the unknown strains ϕ and χ from the equations
$$\phi + \chi = \varpi,$$
$$\phi\chi = \theta.$$

14. Shew that the value of $V(\phi a\chi a + \phi\beta\chi\beta + \phi\gamma\chi\gamma)$ is the same, whatever rectangular unit system is denoted by a, β, γ.

15. Find a system of simple shears whose successive application results in a pure strain.

16. Shew that, if ϕ be self-conjugate, and ξ, η two vectors, the two following equations are consequences one of the other :—

$$\frac{\xi}{S^{\frac{1}{3}} . \xi\phi\xi\phi^2\xi} = \frac{V . \eta\phi\eta}{S^{\frac{2}{3}} . \eta\phi\eta\phi^2\eta},$$

$$\frac{\eta}{S^{\frac{1}{3}} . \eta\phi\eta\phi^2\eta} = \frac{V . \xi\phi\xi}{S^{\frac{2}{3}} . \xi\phi\xi\phi^2\xi}.$$

From either of them we obtain the equation :

$$S\phi\xi\phi\eta = S^{\frac{1}{3}} . \xi\phi\xi\phi^2\xi S^{\frac{1}{3}} . \eta\phi\eta\phi^2\eta.$$

14—2

17. Shew that in general any self-conjugate linear and vector function may be expressed in terms of two given ones, the expression involving terms of the second order.

Shew also that we may write

$$\phi + z = a\left(\varpi + x\right)^2 + b\left(\varpi + x\right)\left(\omega + y\right) + c\left(\omega + y\right)^2,$$

where a, b, c, x, y, z are scalars, and ϖ, ω the given functions. What character of generality is necessary in ϖ and ω? How is the solution affected by non-self-conjugation in one or both?

18. Solve the equations:

(a) $V . \alpha\rho\beta = V . \alpha\gamma\beta,$

(b) $\alpha\rho + \rho\beta = \gamma,$

(c) $\rho + \alpha\rho\beta = \alpha\beta,$

(d) $\alpha\rho\alpha^{-1} + \beta\rho\beta^{-1} = \gamma\rho\gamma^{-1},$

(e) $\alpha\rho\beta\rho = \rho\alpha\rho\beta.$

APPENDIX.

WE have thought it would be acceptable to many students if we should give as an Appendix a brief, and in some cases even a detailed, solution of the most important and most difficult of the ADDITIONAL EXAMPLES. In doing so, we would add as a word of advice, that our solutions be employed simply for the purpose of comparison with those which shall occur to the student himself.

CHAP. II.

Ex. 4. If $AB = a$, $BC = \beta$, $AP = ma$, $AP' = m'a$, $BQ = m\beta$, &c.; then

$$AE = AP + xPQ = AP' + x'P'Q'$$

gives $\quad ma + x\{(1-m)\,a + m\beta\} = m'a + x'\{(1-m')\,a + m'\beta\}$,

whence $\qquad\qquad x = m'$, and $PE = m'PQ$.

Ex. 6. $ABCD$ is a quadrilateral; $AB = a$, $AC = \beta$, $AD = \gamma$, $AP = ma$, $BQ = m\,(\beta - a)$, &c.

The condition $\qquad PQ + RS = 0$

gives $\qquad (1-m)\,a + m\,(\beta - a) + (1-m)\,(\gamma - \beta) - m\gamma = 0$,

or $\qquad\qquad (1-2m)\,(a - \beta + \gamma) = 0;$

an equation which is satisfied either when $1 - 2m = 0$, or when $a - \beta + \gamma = 0$.

The former solution is Ex. 5; the latter gives $ABCD$. a parallelogram.

Ex. 10. Let a, b, c be the points in which the bisectors of the exterior angles at A, B, C meet the opposite sides. Let unit vectors along BC, CA, AB be α, β, γ; then with the usual notation we have

$$a\alpha + b\beta + c\gamma = 0\dots\dots\dots\dots\dots\dots(1).$$

Now
$$A\alpha = x(\beta + \gamma) = -b\beta + y(b\beta + c\gamma)$$

gives
$$x = \frac{bc}{b-c};$$

and
$$A\alpha = \frac{bc}{b-c}(\beta + \gamma).$$

Similarly
$$Bb = \frac{ca}{c-a}(\gamma + \alpha),$$

$$Cc = \frac{ab}{a-b}(\alpha + \beta),$$

therefore
$$Ab = -\frac{bc}{c-a}\beta, \text{ (by 1)},$$

$$Ac = -\frac{bc}{a-b}\gamma.$$

Hence
$$(b-c)A\alpha + (c-a)Ab + (a-b)Ac = 0,$$

and also
$$(b-c) + (c-a) + (a-b) = 0,$$

therefore (Art. 13) a, b, c are in a straight line.

COR.
$$ba : ca :: b-a : c-a.$$

Ex. 12. If the figure of Ex. 11, Art. 23, be supposed to represent a parallelepiped; then, with the notation of that example, the vector from O to the middle point of OG is $\frac{1}{2}(\alpha + \beta + \delta)$, which is the same as the vector to the middle point of AF, viz.

$$\alpha + \frac{1}{2}(\beta + \delta - \alpha).$$

Ex. 13. With the figure and notation of Art. 31, the former part of the enunciation is proved by the equation

$$\frac{a+\beta+\gamma}{4} = \frac{1}{4}\left(\frac{a+\beta+\gamma}{3} + \frac{a+\beta}{3} + \frac{\beta+\gamma}{3} + \frac{\gamma+a}{3}\right).$$

Also, if the edges AB, BC, CA be bisected in c, a, b, the mean point of the tetrahedron $Oabc$ is evidently

$$\frac{1}{4}\left(\frac{a+\beta}{2} + \frac{\beta+\gamma}{2} + \frac{\gamma+a}{2}\right),$$

which proves the latter part of the enunciation.

Ex. 14. Here we have to do with nothing but the triangles on each side of OD.

If $OQ = a$, $QA = pa$, $AP = \beta$, $PD = q\beta$;

$$TO = xOD = TQ - OQ = yQP - OQ$$

gives $$x = \frac{1}{pq-1}.$$

Similarly, if $OS = a'$, $SB = p'a'$, $BR = \beta'$, $RD = q'\beta'$;

$$T'O = x'OD$$

gives $$x' = \frac{1}{p'q'-1}.$$

But the data are $\dfrac{1}{q} = \dfrac{m}{p}$, $p = mq'$; hence

$$pq = p'q', \text{ and } x = x';$$

therefore T' coincides with T.

Ex. 15. If $AB = a$, $AC = \beta$, $MN = pa$, $PQ = q\beta$, $RS = r(\beta-a)$, we shall have, by making $AO = AP + PO = AR + RO$,

$$(1-q)a + (1-p)\beta = ra + (1-p)(\beta-a);$$

therefore $$p + q + r = 2.$$

Ex. 17. Let $RA = a$, $RB = \beta$, $AP = ma$, $AD = pa + q\beta$; then

$$PD = pa + q\beta - ma,$$

and $RS = RP + PS = RQ + QS$ gives

$$(1 + m) \alpha + x (p\alpha + q\beta - m\alpha) = (1 + m) \beta + y (p\alpha + q\beta - m\beta),$$

whence $$x = \frac{1 + m}{m},$$

and $$RS = \frac{1 + m}{m} (p\alpha + q\beta) = \frac{1 + m}{m} AD.$$

[Or thus :

$$\overline{DA} = \alpha, \quad \overline{DC} = \beta,$$

$$\overline{DQ} = m\alpha \quad \therefore \quad \overline{DS} = \frac{1}{2} (\beta + m\alpha),$$

$$\overline{QA} = (1 - m) \alpha; \quad \overline{QR} = \frac{1}{2} (\beta + \overline{1 - m}\,\alpha),$$

$$\overline{DR} = \frac{1}{2} (\beta + \overline{1 + m}\,\alpha),$$

$$\overline{SR} = \overline{DR} - \overline{DS} = \frac{1}{2} \alpha.]$$

CHAP. III.

Ex. 5.   Let $ABCD$ be the quadrilateral; $DA, DB, DC, \alpha, \beta, \gamma$ respectively.

Now $$\beta (\gamma - \alpha) + (\gamma - \alpha) \beta = \gamma (\beta - \alpha) + (\beta - \alpha)\gamma$$
$$+ \alpha (\gamma - \beta) + (\gamma - \beta) \alpha.$$

Taking scalars, and applying 22. 3, there results,

$$S\beta (\gamma - \alpha) = S\gamma (\beta - \alpha) + S\alpha (\gamma - \beta),$$

which is the proposition.

Ex. 6.   If α, β, γ be the vectors OA, OB, OC corresponding to the edges a, b, c; we have

$$V (CA . CB) = V (\alpha - \gamma) (\beta - \gamma)$$
$$= V (\alpha\beta + \beta\gamma + \gamma\alpha)$$
$$= abk + bci + caj,$$

the negative square of which is the proposition given.

Ex. 7. If $S\alpha(\beta-\gamma)=0$ and $S\beta(\alpha-\gamma)=0$, then, by subtraction, will $S\gamma(\alpha-\beta)=0$.

Ex. 8. If $\alpha^2=(\beta-\gamma)^2$, $\beta^2=(\gamma-\alpha)^2$, $\gamma^2=(\alpha-\beta)^2$; then will

$$S\left(\frac{\beta+\gamma}{2}-\frac{\alpha}{2}\right)\left(\frac{\gamma+\alpha}{2}-\frac{\beta}{2}\right)=0, \text{ &c. &c.,}$$

for these are the same equations in another form; and they prove that the corresponding vectors are at right angles to one another.

Ex. 9. If OA, OB, OC, OD are α, β, γ, δ;

triangle DAB : DAC :: tetrahedron $ODAB$: $ODAC$

:: $S\alpha\beta\delta$: $S\alpha\gamma\delta$

:: triangle OAB : OAC,

because the angles which δ makes with the planes OAB, OAC are equal.

Chap. IV.

Ex. 1. Let O be the middle point of the common perpendicular to the two given lines; α, $-\alpha$, the vectors from O to those lines, unit vectors along which are β, γ; ρ the vector to a point P in a line QR which joins the given lines; P being such that $RP=mPQ$; therefore

$$\rho+\alpha-y\gamma=m(\alpha+x\beta-\rho).$$

Now since α is perpendicular to both β and γ, the equation gives $(1+m)S\alpha\rho=(m-1)\alpha^2$; a plane.

Ex. 2. Retaining what is necessary of the notation of the last example, let $OS=\delta$.

If PR perpendicular on γ meet β in Q, we have

$$-\alpha+y\gamma+RP=\rho, \text{ which gives } y\gamma^2=S\gamma\rho;$$
$$RQ=2\alpha+x\beta-y\gamma, \text{ which gives } y\gamma^2=xS\beta\gamma;$$

and $SP^2 = e^2PQ^2$ gives

$$(\rho - \delta)^2 = e^2 (\alpha + x\beta - \rho)^2$$
$$= e^2 \left(\alpha + \frac{S\gamma\rho}{S\beta\gamma} \beta - \rho \right)^2,$$

which being of the second degree in ρ shews that the locus is a surface of the second order. See Chap. VI.

Ex. 3.   The equation of the plane is

$$S\gamma\rho = a,$$

which, being substituted in the equation of the surface, gives what is obviously the equation of a circle.

Ex. 4.   With the notation of Ex. 1, let δ, δ' be the perpendiculars on the lines,

then $\rho + \delta = \alpha + x\beta$ gives $V\beta\delta = -V\beta(\rho - \alpha)$,

and the condition given may be written

$$V^2\beta\delta = e^2V^2\gamma\delta';$$
$$\therefore\ V^2\beta(\rho - \alpha) = e^2V^2\gamma(\rho + \alpha).$$

Now (22. 9)

$$V^2\beta(\rho - \alpha) = -\beta^2(\rho - \alpha)^2 + S^2\beta(\rho - \alpha),$$

whence $\rho^2 - 2S\alpha\rho + \alpha^2 + S^2\beta\rho = e^2(\rho^2 + 2S\alpha\rho + \alpha^2 + S^2\gamma\rho)$,

a surface of the second order.

Ex. 6. $S\rho(\beta + \gamma) = c$, a plane perpendicular to the line which bisects the angle which parallels to the given lines drawn through O make with one another.

Ex. 7. α, β the vectors to the given points A, B,

$$S\gamma\rho = a,\quad S\delta\rho = b$$

the equations of the planes, γ, δ being unit vectors.
$x\gamma$, $y\delta$ the vector perpendiculars from A on the planes, then

$$x = S\alpha\gamma - a,\quad y = S\alpha\delta - b,$$
$$\therefore\ x + y = S\alpha(\gamma + \delta) - (a + b)\ldots\ldots\ldots\ldots\ldots(1).$$

Hence by the question

$$Sa\,(\gamma + \delta) = S\beta\,(\gamma + \delta)$$

or
$$S\,(\beta - a)\,(\gamma + \delta) = 0\ldots\ldots\ldots\ldots\ldots\ldots(2).$$

Now equation (1) will give the sum of the perpendiculars on the planes from any other point in the line AB by simply writing $a + z\,(\beta - a)$ in place of a; and from equation (2) this will produce no change.

Ex. 8. If β' be the vector to C, equation (2) of the last example gives

$$S\,(\beta - a)\,(\gamma + \delta) = 0, \quad S\,(\beta' - a)\,(\gamma + \delta) = 0.$$

Now the sum of the perpendiculars from any other point in the plane will be found from equation (1) by writing

$$a + z\,(\beta - a) + z'\,(\beta' - a)$$

in place of a. Hence the proposition.

Ex. 10. Tait's *Quaternions*, Art. 213.

Ex. 11. Let a, β, γ, δ be the vectors OA, OB, OC, OD; then (34. 5, Cor.)

$$\delta = S.\,a\beta\gamma\,.\,(Va\beta + V\beta\gamma + V\gamma a)^{-1}$$

$$= \frac{abc\,(bci + caj + abk)}{(ab)^2 + (bc)^2 + (ca)^2}\ \ldots\ldots\ldots\ldots\ldots\ldots\ldots(1).$$

Now

triangle ABD : triangle ABC

 :: tetrahedron $OABD$: tetrahedron $OABC$

  :: $S.\,a\beta\delta$: $S.\,a\beta\gamma$

  :: $S.\,abij\delta$: $S.\,abcijk$

  :: $(ab)^2$: $(ab)^2 + (bc)^2 + (ca)^2$

  :: (triangle $AOB)^2$: (triangle $ABC)^2$.

(Chap. III., Additional Ex. 6.)

Ex. 12. This is merely the equation

$$\rho = at + \frac{\beta}{t},$$

with t eliminated by taking the product of $V\alpha\rho$, $V\beta\rho$. (See 55. 3.)

CHAP. V.

Ex. 3. Let a, a' be the radii of the circles; α, ρ the vectors from the centre of one of them to that of the other, and to the point whose locus is required; then

$$\frac{T\rho}{a} = \frac{T(\rho - \alpha)}{a'}.$$

Ex. 7. This is the polar reciprocal of Ex. 3, Art. 40.

Ex. 8. Let A be the origin, $AB = \beta$, $AC = \gamma$, the vector to the centre α: then

$$-V(AB . BC . CA) = V . \beta(\gamma - \beta)\gamma$$
$$= \gamma^2\beta - \beta^2\gamma$$
$$= 2\beta S\alpha\gamma - 2\gamma S\alpha\beta \text{ from the circle;}$$
$$\therefore S . \alpha V(AB . BC . CA) = 0.$$

Ex. 9. Tait, Art. 222.

Ex. 10. Tait, Art. 221.

Ex. 11. Tait, Art. 223.

Ex. 12. Tait, Art. 232.

CHAP. VI.

Ex. 1. Let δ be the vector to the given point, π the vector to the point of bisection of a chord, β a vector parallel to the chord, all measured from the centre; then

$$\delta = \pi + x\beta,$$
$$S\pi\phi\delta = S\pi\phi\pi \dots\dots\dots\dots\dots\dots(48);$$

from which by making

$$\pi = \rho + \frac{1}{2}\,\delta,$$

we get $$S\rho\phi\rho = \frac{1}{4}\,S\delta\phi\delta,$$

an ellipse whose centre is at the point of bisection of the line which joins the given point with the centre of the given ellipse.

Ex. 2. Let $2b$ be the shortest distance between the given lines; θ their angle of inclination; $2a$ the line of constant length; then as in Ex. 2, Chap. IV.,

$$-4a^2 = (2a + x\beta - y\gamma)^2,$$
$$2\rho = x\beta + y\gamma\,;$$

the former gives

$$x^2 + y^2 - 2xy \cos\theta = 4\,(a^2 - b^2)\ldots\ldots\ldots\ldots\ldots(1),$$

the latter

$$4\rho = (x + y)\,(\beta + \gamma) + (x - y)\,(\beta - \gamma),$$

which, since $\beta + \gamma$, $\beta - \gamma$ are vectors bisecting the angles between the lines and therefore at right angles to one another, is an equation of the form of that in Art. 55. 2; whilst equation (1) satisfies the condition

$$(x + y)^2 + m\,(x - y)^2 = c,$$

which is requisite for an ellipse.

Ex. 3. Let a be a vector semi-diameter, parallel to a chord through O; δ the vector to O: then

$$\rho = \delta + xa$$

gives $$S\delta\phi\delta + 2xS\delta\phi a + x^2 Sa\phi a = 1,$$

which, since $$Sa\phi a = 1,$$

shews that the product of the two values of x is constant; hence the rectangle by the segments of the chord varies as a^2, which is the proposition.

Ex. 4. With the usual notation, let CE, CE' be semi-diameters parallel to DP, $D'P$, and let their vectors be $m(a - \beta)$, $n(a + \beta)$; then since P, D, E, E' are points in the ellipse,

$$m^2 S(a - \beta) \phi(a - \beta) = 1,$$

$$\therefore \ 2m^2 = 1. \ \text{Similarly } 2n^2 = 1, \ m = n,$$

$$\text{and} \quad DP \ : \ D'P \ :: \ T(a - \beta) \ : \ T(a + \beta)$$

$$:: \ Tm(a - \beta) \ : \ Tn(a + \beta)$$

$$:: \ CE \ : \ CE'.$$

Cor. Since $m = \dfrac{1}{\sqrt{2}}$, $CE : DP :: 1 : \sqrt{2}$.

Ex. 5. Put na', $n\rho'$ in place of a, ρ in equation (1), Art. 43.

Ex. 6, 7. With everything as in Ex. 4, CE, CE' being now semi-diameters in the direction of diagonals of the parallelogram,

$$SCE\phi CE' = \frac{1}{2} S(a - \beta) \phi(a + \beta)$$

$$= 0 ;$$

hence CE, CE' are conjugate.

Ex. 8. $S(a + \beta) \phi(a + \beta) = 2$ gives an ellipse, whose equation is

$$S\rho\phi'\rho = 1, \ \text{where} \ \phi' = \frac{\phi}{2} ;$$

hence the diameters of the locus are to those of the given ellipse

$$:: \ \sqrt{2} \ : \ 1.$$

Ex. 9. If γ be a unit vector to which the lines are parallel, ρ, ρ' points in which the lines cut the ellipse,

$$\rho = ai + m\gamma, \ \rho' = bj + n\gamma,$$

and $\qquad\qquad\qquad S\rho\phi\rho = 1$ gives

$$2aSi\phi\gamma + mS\gamma\phi\gamma = 0 \rbrace$$
Similarly $\qquad 2bSj\phi\gamma + nS\gamma\phi\gamma = 0 \rbrace \cdots\cdots\cdots\cdots(1).$

Now $\qquad S\rho\phi\rho' = an\,Si\phi\gamma + bm\,Sj\phi\gamma + mn\,S\gamma\phi\gamma$

$\qquad\qquad\qquad = 0$, by equations (1);

$\qquad\qquad \therefore\ \rho,\ \rho'$ are conjugate.

COR. The same demonstration applies when the diameters from whose extremities parallels are drawn, are any conjugate diameters whatever, i, j being parallel to those diameters.

Ex. 10. Let CP, CP' be any two semi-diameters, their vectors being $a,\ a'$; PQ the semi-ordinate to CP'; $CQ = na'$; then

$$S\,(PQ \cdot \phi a') = 0$$

gives $\qquad\qquad S\,(a - na')\,\phi a' = 0,$

$$\therefore\ n = Sa\phi a'.$$

Now the area of the triangle QCP is proportional to

$$V\,(CP \cdot CQ),$$

i. e. to $nVaa'$ or to

$$Sa\phi a' \cdot Vaa',$$

which, being symmetrical in $a,\ a'$, proves the proposition.

Ex. 11. If the tangent at P' meet CP produced in T,

$$CT = ma\ ;$$

then, since $P'T$ is perpendicular to $\phi a'$,

$$S\,(CT - a')\,\phi a' = 0.$$

$$\therefore\ m = \frac{1}{Sa\phi a'}\,;$$

and area $P'CT$ is proportional to $V\,(CP' \cdot CT)$, i.e. to $\dfrac{Vaa'}{Sa\phi a'}$, which is symmetrical in $a,\ a'$.

Ex. 12. Let $a,\ \beta$ be the vector semi-diameters of the larger ellipse; C the centre; O the centre of the smaller ellipse, whose equation is

$$S\rho\phi\rho = c,$$

γ a vector along PQR; then

$$OQ = \frac{\alpha + \beta}{2} + x\gamma \, ;$$

$$\therefore \ S\left(\frac{\alpha + \beta}{2} + x\gamma\right)\phi\left(\frac{\alpha + \beta}{2} + x\gamma\right) = c$$

$$= S\frac{\alpha + \beta}{2}\phi\frac{\alpha + \beta}{2} \, ,$$

$$\therefore \ x = -\frac{S\left(\alpha + \beta\right)\phi\gamma}{S\gamma\phi\gamma} \, ;$$

and since

$$CQ = \alpha + \beta + x\gamma,$$

$$S\left(CQ\phi\gamma\right) = 0 \, ;$$

hence PR is conjugate to CQ, and therefore bisected at Q.

Ex. 13.   This is simply a combination of 49. 2 and 49. 1.

CHAP. VII.

Ex. 3.   The equation of the circle is

$$\left(\rho - \frac{a}{2}\right)^2 = \frac{9}{16}a^2,$$

which by 52. 1 gives

$$\left(a^2 - Sa\rho\right)^2 - a^2 Sa\rho = \frac{5}{16}a^4,$$

$$\therefore \ Sa\rho = \frac{a^2}{4} \, ,$$

which (52. 11) is the proposition.

Ex. 5.   If O be the centre of the circle, Q a point at which it meets the tangent at A; then, with the notation of 55. 1,

$$QO^2 = \{a\alpha + \frac{1}{2}\left(\rho - a\alpha\right) - z\beta\}^2 = \frac{1}{4}\left(\rho - a\alpha\right)^2,$$

$$\therefore \ z^2\beta^2 - zS\beta\rho + a_\prime Sa\rho = 0,$$

$$\text{i. e. } \ z^2 - zy + \frac{y^2}{4} = 0,$$

which gives two equal values of z; hence the proposition.

Ex. 6. With any point as origin, let β, γ be the vectors to the two given points, π the vector to the focus of one of the parabolas. Write $a\alpha$ in place of α in equation (1), Art. 52, α being a unit vector;

then
$$- (\beta - \pi)^2 = \{a + S\alpha\,(\beta - \pi)\}^2 \ldots\ldots\ldots\ldots\ldots\ldots(1)$$
$$- (\gamma - \pi)^2 = \{a + S\alpha\,(\gamma - \pi)\}^2,$$

whence, by subtraction,

$$\beta^2 - \gamma^2 - 2S\pi\,(\beta - \gamma) = - S\alpha\,(\beta - \gamma)\,\{2a + S\alpha\,(\beta - \gamma) - 2S\alpha\pi\},$$

which gives a by a simple equation in π; and then equation (1) becomes a quadratic in π.

Ex. 8. If two tangents meet at T, it is easy, as in Ex. 5, Art. 55, with the notation available for the focus, to find

$$ST = \frac{yy'}{4a}\,\alpha + \frac{y + y'}{2}\,\beta - a\alpha,$$

$$ST' = \frac{yy''}{4a}\,\alpha + \frac{y + y''}{2}\,\beta - a\alpha,$$

and $S\,(ST\,.\,ST') = 0$ will follow at once, from the fact that

$$y'y'' + 4a^2 = 0.$$

Ex. 9. Let P be the point of contact, PQ the chord, TEF the line parallel to the axis cutting the curve in Ej; E the origin;

$$EP = \frac{t^2}{2}\,\alpha + t\beta, \quad ET = -\frac{t^2}{2}\,\alpha,$$

$$EP = EF + FP = y\alpha + z\left\{\frac{t^2 - t'^2}{2}\,\alpha + (t - t')\beta\right\};$$

whence
$$z = \frac{t}{t - t'}, \quad y = -\frac{tt'}{2}.$$

$$\therefore PF \,:\, FQ \,::\, t \,:\, t'$$
$$:: \frac{t^2}{2} \,:\, \frac{tt'}{2}$$
$$:: TE \,:\, EF.$$

Ex. 10. This is evident from equation (1), Art. 52.

T. Q.   15

Ex. 11. With the notation of Art. 52, let

$$SQ = xPS = -x\rho, \quad Ay = -x'AP = x'\left(\frac{a}{2} - \rho\right),$$

$$\therefore \quad x'(a - 2\rho) = a + \delta,$$

$$x'(a^2 - 2Sa\rho) = a^2.$$

But ρ, $-x\rho$ being vectors to the parabola, equation (1), Art. 52, gives

$$x^2(a^2 - Sa\rho)^2 = (a^2 + xSa\rho)^2,$$

$$\therefore \quad x(a^2 - Sa\rho) = a^2 + xSa\rho,$$

$$x(a^2 - 2Sa\rho) = a^2,$$

$$\therefore \quad x = x',$$

and the proposition is true (Euc. VI. 2).

Ex. 14. Tait, Art. 43, Cor. 2.

Ex. 15.

$$CP = at + \frac{\beta}{t} \text{ gives } CT = 2at,$$

$$CQ = 2at + x\beta = at' + \frac{\beta}{t'} = 2at + \frac{\beta}{2t},$$

$$\therefore \quad PQ = at - \frac{\beta}{2t},$$

so that the equation of $RQPR'$ is

$$\rho = at + \frac{\beta}{t} + x\left(at - \frac{\beta}{2t}\right),$$

whence for R and R' the values of x are 2 and -1; therefore

$$CR = 3at, \quad CR' = \frac{3}{2}\frac{\beta}{t},$$

$$QR = at - \frac{\beta}{2t} = PQ = \frac{1}{3}RR'.$$

Ex. 16.   If $CR = a\alpha$; $\alpha + m\beta$, $\alpha - m\beta$ vectors parallel to the given conjugate diameters,

$$CP = a\alpha + x\left(\alpha + m\beta\right) = \alpha t + \frac{\beta}{t},$$

$$CD = a\alpha + x'\left(\alpha - m\beta\right) = \alpha t' - \frac{\beta}{t'},$$

give $t = t'$; therefore CP, CD are conjugate.

Ex. 18.   Adopting the figure and notation of Ex. 2 of the hyperbola, Art. 55, we have

$$CR = 2Xt\alpha, \quad Cr = 2X\frac{\beta}{t};$$

therefore $\qquad QR = (X - Y)\left(t\alpha - \frac{\beta}{t}\right),$

$$rQ = (X + Y)\left(t\alpha - \frac{\beta}{t}\right),$$

and $\qquad rQ.QR = (X^2 - Y^2)\left(t\alpha - \frac{\beta}{t}\right)^2$

$$= PO^2, \text{ since } X^2 - Y^2 = 1.$$

As an example of combining not merely the forms but the results of the Cartesian Geometry with Quaternions, we will add one more example.

If CP, CD; CP', CD' be two pairs of conjugate semi-diameters of an ellipse, PD' will be parallel to $P'D$.

Let CP, CP' be denoted, as in Art. 55. 2, by $x\alpha + y\beta$, $x'\alpha + y'\beta$ respectively; then CD, CD' will be represented by

$$-\frac{a}{b}y\alpha + \frac{b}{a}x\beta, \quad -\frac{a}{b}y'\alpha + \frac{b}{a}x'\beta,$$

with the conditions

$$a^2y^2 + b^2x^2 = a^2b^2, \quad a^2y'^2 + b^2x'^2 = a^2b^2\ldots\ldots\ldots(1).$$

Now $\qquad$ vector $D'P = \left(x + \frac{a}{b}y'\right)\alpha + \left(y - \frac{b}{a}x'\right)\beta,$

$$DP' = \left(x' + \frac{a}{b}y\right)\alpha + \left(y' - \frac{b}{a}x\right)\beta.$$

But equations (1) give, by subtraction,

$$x + \frac{a}{b}y' \;:\; y - \frac{b}{a}x' \;::\; x' + \frac{a}{b}y \;:\; y' - \frac{b}{a}x$$

therefore $D'P$ is a multiple of DP' and consequently parallel to it.

Cor. $\qquad\qquad PD' \;:\; P'D \;::\; ay' + bx \;:\; ay + bx'.$

Chap. VIII.

Ex. 1. With the notation of Additional Ex. 1, Chap. IV., the perpendiculars are

$$\rho - a - x\beta, \quad \rho + a - y\gamma,$$

so that $\qquad\qquad S\beta\rho = x\beta^2, \quad S\gamma\rho = y\gamma^2;$

and by the question,

$$(\rho - a - \beta^{-1}S\beta\rho)^2 = e^2(\rho + a - \gamma^{-1}S\gamma\rho)^2,$$

a surface of the second order in ρ.

Ex. 3. The equations $S\rho\phi\rho = 1$, $S\pi\phi\rho = 1$, with the condition $\pi = x\phi\rho$, give

$$\frac{1}{x^2}S\pi\phi^{-1}\pi = 1, \quad \frac{\pi^2}{x} = 1 \text{ respectively,}$$

therefore $\qquad\qquad S\pi\phi^{-1}\pi = \pi^4,$

whence the Cartesian equation.

Ex. 4. If a, β, γ are the vector radii,

$$\frac{Sa\phi a}{(Ta)^2} = \frac{(SiUa)^2}{a^2} + \frac{(SjUa)^2}{b^2} + \frac{(SkUa)^2}{c^2},$$

$$\&c. = \&c.$$

Adding and observing that $Sa\phi a = 1$, &c., there results

$$\frac{1}{(Ta)^2} + \frac{1}{(T\beta)^2} + \frac{1}{(T\gamma)^2} = \frac{1}{a^2} + \frac{1}{b^2} + \frac{1}{c^2}.$$

Ex. 5. As in Ex. 8, Art. 64,

$$\frac{1}{\overline{Oy_1}^2} = -(\phi a)^2,$$

and if vector $OQ_1 = x\phi a$, the ellipsoid gives

$$x^2 S\phi a \phi^2 a = 1.$$

Now
$$\frac{1}{\overline{Oy_1}^2 . \overline{OQ_1}^2} = \frac{1}{x^2} = S\phi a \phi^2 a$$

$$= \frac{(Sia)^2}{a^6} + \frac{(Sja)^2}{b^6} + \frac{(Ska)^2}{c^6},$$

and, since

$$(Sia)^2 + (Si\beta)^2 + (Si\gamma)^2 = a^2$$

(Ex. 7, Art. 64), the result required is obtained by simply adding.

Ex. 6. Let pk be the vector distance from the origin, of the plane parallel to xy, π a point in it; then $Sk (\pi - pk) = 0$ gives $S\pi k = $ const.

Now $S\rho\phi\pi = 1$ is the equation of the plane of contact, and if zk be the point in which this plane cuts the axis of z, $zSk\phi\pi = 1$, i.e. $zS\pi\phi k = 1$, gives z.

Now ϕk is a multiple of k, and since $S\pi k$ is constant, z is constant.

Ex. 7. The equations of the ellipsoids

$$S\rho\phi\rho = 1, \quad S(\rho - a)\phi(\rho - a) = 1,$$

give $S\rho\phi a = $ const. as the plane of contact.

Ex. 8. If pa be the vector to the point in the line OA; the equation of its polar plane is $Spa\phi\rho = 1$; and the square of the reciprocal of the perpendicular from the centre on this plane is $-p^2(\phi a)^2$. Hence the conclusion by Ex. 8, Art. 64.

Ex. 9. Let ρ be the vector to P; a, β, γ vector radii parallel to the chords; then

$$\rho + xa, \quad \rho + y\beta, \quad \rho + z\gamma,$$

will be the vectors to A, B, C; and since P, A, B, C are points in the ellipsoid

$$S\rho\phi\rho = 1, \quad 2S\rho\phi\alpha + x = 0, \quad 2S\rho\phi\beta + y = 0,$$
$$2S\rho\phi\gamma + z = 0.$$

The equation of the plane ABC is (34. 5)

$$S \cdot (\pi - \rho)(xy\alpha\beta + yz\beta\gamma + zx\gamma\alpha) = xyzS \cdot \alpha\beta\gamma,$$

and since α, β, γ are at right angles to one another,

$$\alpha\beta = -\frac{\gamma}{(T\gamma)^2} S \cdot \alpha\beta\gamma, \quad \&c.$$

therefore the equation of the plane ABC becomes

$$S \cdot (\pi - \rho)\left\{\frac{1}{(T\gamma)^2} \cdot \frac{\gamma}{S\rho\phi\gamma} + \frac{1}{(T\alpha)^2} \cdot \frac{\alpha}{S\rho\phi\alpha} + \frac{1}{(T\beta)^2} \cdot \frac{\beta}{S\rho\phi\beta}\right\} = 2,$$

which is satisfied by

$$\pi - \rho = m\phi\rho,$$

where

$$m\left\{\frac{1}{(T\alpha)^2} + \frac{1}{(T\beta)^2} + \frac{1}{(T\gamma)^2}\right\} = 2 \; ;$$

and therefore Ex. 4 above gives

$$m = \frac{2}{\dfrac{1}{a^2} + \dfrac{1}{b^2} + \dfrac{1}{c^2}} \; .$$

Chap. IX.

Ex. 2 and 3.   Employ formula 11.

Ex. 5.   Since

$$\alpha^2\beta^2\gamma^2 = \alpha\beta\gamma \cdot \gamma\beta\alpha,$$

formulæ 4 and 6 give the required result.

Ex. 6.   Apply formula 10 to Ex. 5.

Ex. 8. $(\alpha\beta\gamma)^2 = \alpha\beta\gamma \cdot \alpha\beta\gamma = \alpha\beta\gamma \, (S \cdot \alpha\beta\gamma + V \cdot \alpha\beta\gamma)$

$\qquad\qquad = \alpha\beta\gamma \, (S \cdot \alpha\beta\gamma + V \cdot \gamma\beta\alpha)$

$\qquad\qquad = \alpha\beta\gamma \, (\gamma\beta\alpha + 2S \cdot \alpha\beta\gamma)$

$\qquad\qquad = \alpha^2\beta^2\gamma^2 + 2\alpha\beta\gamma S \cdot \alpha\beta\gamma.$

Ex. 9. Formula 10 gives the vector of the product of three vectors α, β, γ, under the form $\alpha' - \beta' + \gamma'$ where $\alpha' = \alpha S\beta\gamma$, &c.

Hence the required scalar may be written

$$S \cdot (\alpha' - \beta' + \gamma')(\alpha' + \beta' - \gamma')(-\alpha' + \beta' + \gamma') ;$$

and as the scalar part of this product is that which involves all of the three vectors α', β', γ' we have exactly as in the demonstration of formula 5,

$$
\begin{aligned}
&S(V\alpha\beta\gamma\, V\beta\gamma\alpha\, V\gamma\alpha\beta) \\
&= S \cdot \begin{vmatrix} \alpha', & -\beta', & \gamma' \\ \alpha', & \beta', & -\gamma' \\ -\alpha', & \beta', & \gamma' \end{vmatrix} \\
&= 4S \cdot \alpha'\beta'\gamma'.
\end{aligned}
$$

10. The scalar part, by formula 16, is reduced to

$$S\alpha\delta S\beta\gamma - S\alpha\gamma S\beta\delta - S\alpha\delta S\beta\gamma + S\alpha\beta S\gamma\delta + S\alpha\gamma S\beta\gamma - S\alpha\beta S\gamma\delta,$$

which is identically 0.

The vector part, by formula 12, is

$$\alpha S \cdot \gamma\delta\beta - \beta S \cdot \gamma\delta\alpha + \alpha S \cdot \delta\beta\gamma - \gamma S \cdot \delta\beta\alpha + \alpha S \cdot \beta\gamma\delta - \delta S \cdot \beta\gamma\alpha,$$

which, by formula 13, reduces to

$$2\alpha S \cdot \beta\gamma\delta.$$

12. If, for brevity, we denote $S \cdot \alpha\beta\gamma$, $V \cdot \alpha\beta\gamma$ respectively by S and V, we have, by formula 7,

$$
\begin{aligned}
&2\alpha^2\beta^2\gamma^2 + \alpha^2(\beta\gamma)^2 + \beta^2(\alpha\gamma)^2 + \gamma^2(\alpha\beta)^2 - (\alpha\beta\gamma)^2 \\
&= 2\alpha\beta \cdot \gamma\beta\alpha + \beta\gamma\alpha \cdot \alpha\beta\gamma + \alpha\gamma\beta \cdot \beta\alpha\gamma + \alpha\beta\gamma \cdot \gamma\alpha\beta - (\alpha\beta\gamma)^2 \\
&= 2(S + V)(- S + V) + (S - V + 2\alpha S\beta\gamma)(S + V) \\
&\quad + (- S - V + 2\alpha S\beta\gamma)(- S - V + 2\gamma S\alpha\beta) \\
&\quad + (S + V)(S - V + 2\gamma S\alpha\beta) - (S + V)^2 \\
&= 4\alpha\gamma S\alpha\beta S\beta\gamma.
\end{aligned}
$$

The student is recommended to verify a few examples such as the above, by putting

$$\alpha = i, \quad \beta = ai + bj + ck, \quad \gamma = a'i + b'j + c'k,$$

with the conditions

$$a^2 + b^2 + c^2 = 1, \quad a'^2 + b'^2 + c'^2 = 1.$$

The quaternion equality will then reduce itself to four algebraic equalities, one of which is obvious, and the others are

$$p^2 + r^2 - a'^2 - a^2 + 2aa'm = 0,$$
$$pq - mr + a'c' + ac - 2ac'm = 0,$$
$$qr + mp + a'b' + ab - 2ab'm = 0,$$

where
$$m = aa' + bb' + cc', \quad p = ab' - a'b',$$
$$q = bc' - b'c, \quad r = ca' - c'a.$$

Ex. 13.

$$S. (a - \delta)(\beta - \delta)(\gamma - \delta) = S. a\beta\gamma - S. \beta\gamma\delta + S. \gamma\delta a - S. \delta a\beta.$$

Ex. 14.   By 34. 8, we have

$$-\frac{a}{d} = \frac{S. \delta\beta\gamma}{S. a\beta\gamma} = \pm \frac{BCD}{ABC};$$

therefore the same Article gives

$$\pm a . BCD \pm \beta . CDA \pm \gamma . DAB \pm \delta . ABC = 0;$$

and since the scalar of the product of this vector by the vector perpendicular to the plane in which A, B, C, D lie gives the right-hand side of Ex. 13, we obtain

$$a . BCD - \beta . CDA + \gamma . DAB - \delta . ABC = 0.$$

CAMBRIDGE : PRINTED BY C. J. CLAY, M.A. AT THE UNIVERSITY PRESS.

July 1890.

A Catalogue

OF

Educational Books

PUBLISHED BY

Macmillan & Co.

BEDFORD STREET, STRAND, LONDON

CONTENTS

A

CLASSICS.

Elementary Classics; Classical Series; Classical Library, (1) Texts, (2) Translations; Grammar, Composition, and Philology; Antiquities, Ancient History, and Philosophy.

ELEMENTARY CLASSICS.

18mo, Eighteenpence each.

The following contain Introductions, Notes, and Vocabularies, and in some cases **Exercises.**

ACCIDENCE, LATIN, AND EXERCISES ARRANGED FOR BEGINNERS.—By W. WELCH, M.A., and C. G. DUFFIELD, M.A.

AESCHYLUS.—PROMETHEUS VINCTUS. By Rev. H. M. STEPHENSON, M.A.

ARRIAN.—SELECTIONS. With Exercises. By Rev. JOHN BOND, M.A., and Rev. A. S. WALPOLE, M.A.

AULUS GELLIUS, STORIES FROM.—Adapted for Beginners. With Exercises. By Rev. G. H. NALL, M.A., Assistant Master at Westminster.

CÆSAR.—THE HELVETIAN WAR. Being Selections from Book I. of The Gallic War. Adapted for Beginners. With Exercises. By W. WELCH, M.A., and C. G. DUFFIELD, M.A.

THE INVASION OF BRITAIN. Being Selections from Books IV. and V. of The Gallic War. Adapted for Beginners. With Exercises. By W. WELCH, M.A., and C. G. DUFFIELD, M.A.

THE GALLIC WAR. BOOK I. By Rev. A. S. WALPOLE, MA.

BOOKS II. AND III. By the Rev. W. G. RUTHERFORD, M.A., LL.D.

BOOK IV. By CLEMENT BRYANS, M.A., Assistant Master at Dulwich College.

BOOK V. By C. COLBECK, M.A., Assistant Master at Harrow.

BOOK VI. By the same Editor.

SCENES FROM BOOKS V. AND VI. By the same Editor.

BOOK VII. By Rev. J. BOND, M.A., and Rev. A. S. WALPOLE, M.A.

CICERO.—DE SENECTUTE. By E. S. SHUCKBURGH, M.A.

DE AMICITIA. By the same Editor.

STORIES OF ROMAN HISTORY. Adapted for Beginners. With Exercises. By Rev. G. E. JEANS, M.A., and A. V. JONES, M.A.

EURIPIDES.—ALCESTIS. By M. A. BAYFIELD, M.A.

MEDEA. By the same Editor.   [In the Press

HECUBA. By Rev. J. BOND, M.A., and Rev. A. S. WALPOLE, M.A.

EUTROPIUS.—Adapted for Beginners. With Exercises. By W. WELCH, M.A. and C. G. DUFFIELD, M.A.

HOMER.—ILIAD. BOOK I. By Rev. J. BOND, M.A., and Rev. A. S. WALPOLE, M.A.

BOOK XVIII. By S. R. JAMES, M.A., Assistant-Master at Eton.

ODYSSEY. BOOK I. By Rev. J. BOND, M.A., and Rev. A. S. WALPOLE, M.A.

HORACE.—ODES. BOOKS I.—IV. By T. E. PAGE, M.A., Assistant Master at the Charterhouse. Each 1s. 6d.

LIVY.—BOOK I. By H. M. Stephenson, M.A.
  BOOK XXI. Adapted from Mr. Capes's Edition. By J. E. Melhuish, M.A.
  BOOK XXII. By the same. [*Shortly.*
  THE HANNIBALIAN WAR. Being part of the XXI. and XXII. BOOKS OF LIVY adapted for Beginners. By G. C. Macaulay, M.A.
  THE SIEGE OF SYRACUSE. Being part of the XXIV. and XXV. BOOKS OF LIVY, adapted for Beginners. With Exercises. By G. Richards, M.A., and Rev. A. S. Walpole, M.A.
  LEGENDS OF ANCIENT ROME. Adapted for Beginners. With Exercises. By H. Wilkinson, M.A.
LUCIAN.—EXTRACTS FROM LUCIAN. With Exercises. By Rev. J. Bond, M.A., and Rev. A. S. Walpole, M.A.
NEPOS.—SELECTIONS ILLUSTRATIVE OF GREEK AND ROMAN HISTORY. With Exercises. By G. S. Farnell, M.A.
OVID.—SELECTIONS. By E. S. Shuckburgh, M.A.
  EASY SELECTIONS FROM OVID IN ELEGIAC VERSE. With Exercises. By H. Wilkinson, M.A.
  STORIES FROM THE METAMORPHOSES. With Exercises. By Rev. J. Bond, M.A., and Rev. A. S. Walpole, M.A.
PHÆDRUS — SELECT FABLES. Adapted for Beginners. With Exercises. By Rev. A. S. Walpole, M.A.
THUCYDIDES.—THE RISE OF THE ATHENIAN EMPIRE. BOOK I. Chs. 89-117 and 228-238. With Exercises. By F. H. Colson, M.A.
VIRGIL.—SELECTIONS. By E. S. Shuckburgh, M.A.
  GEORGICS. BOOK I. By T. E. Page, M.A.
  BOOK II. By Rev. J. H. Skrine, M.A.
  ÆNEID. BOOK I. By Rev. A. S. Walpole, M.A.
  BOOK II. By T. E. Page, M.A.
  BOOK III. By T. E. Page, M.A.
  BOOK IV. By Rev. H. M. Stephenson, M.A.
  BOOK V. By Rev. A. Calvert, M.A.
  BOOK VI. By T. E. Page, M.A.
  BOOK VII. By Rev. A. Calvert, M.A.
  BOOK VIII. By Rev. A. Calvert, M.A.   [*In preparation.*
  BOOK IX. By Rev. H. M. Stephenson, M.A.
  BOOK X. By S. G. Owen, M.A.   [*In preparation.*
XENOPHON.—ANABASIS. BOOK I. By Rev. A. S. Walpole, M.A.
  BOOK I. With Exercises. By E. A. Wells, M.A.
  BOOK II. By Rev. A. S. Walpole, M.A.
  BOOK III. By Rev. G. H. Nall.   [*In preparation.*
  SELECTIONS FROM BOOK IV. With Exercises. By Rev. E. D. Stone, M.A.
  BOOK IV. By the same Editor.   [*In preparation.*
  SELECTIONS FROM THE CYROPÆDIA. With Exercises. By A. H. Cooke, M.A., Fellow and Lecturer of King's College, Cambridge.

The following contain Introductions and Notes, but no **Vocabulary** :—

CICERO.—SELECT LETTERS. By Rev. G. E. Jeans, M.A.
HERODOTUS.—SELECTIONS FROM BOOKS VII. and VIII. THE EXPEDITION OF XERXES. By A. H. Cooke, M.A.
HORACE.—SELECTIONS FROM THE SATIRES AND EPISTLES. By Rev. W. J. V. Baker, M.A.
  SELECT EPODES AND ARS POETICA. By H. A. Dalton, M.A., Assistant Master at Winchester.
PLATO.—EUTHYPHRO AND MENEXENUS. By C. E. Graves, M.A., Classical Lecturer at St. John's College, Cambridge.
TERENCE.—SCENES FROM THE ANDRIA. By F. W. Cornish, M.A., Assistant Master at Eton

THE GREEK ELEGIAC POETS.—FROM CALLINUS TO CALLIMACHUS. Selected by Rev. HERBERT KYNASTON, D.D.

THUCYDIDES.—BOOK IV. CHS. 1-41. THE CAPTURE OF SPHACTERIA. By C. E. GRAVES, M.A.

CLASSICAL SERIES
FOR COLLEGES AND SCHOOLS.
Fcap. 8vo.

ÆSCHINES.—IN CTESIPHONTEM. By Rev. T. GWATKIN, M.A., and E. S. SHUCKBURGH, M.A. *[In the Press.*

ÆSCHYLUS.—PERSÆ. By A. O. PRICKARD, M.A., Fellow and Tutor of New College, Oxford. With Map. 3s. 6d.

SEVEN AGAINST THEBES. SCHOOL EDITION. By A. W. VERRALL, Litt.D., Fellow of Trinity College, Cambridge, and M. A. BAYFIELD, M.A., Head-master's Assistant at Malvern College. 3s. 6d.

ANDOCIDES.—DE MYSTERIIS. By W. J. HICKIE, M.A. 2s. 6d.

ATTIC ORATORS.—Selections from ANTIPHON, ANDOCIDES, LYSIAS, ISO-CRATES, AND ISAEUS. By R. C. JEBB, Litt.D., Regius Professor of Greek in the University of Cambridge. 6s.

CÆSAR.—THE GALLIC WAR. By Rev. JOHN BOND, M.A., and Rev. A. S. WALPOLE, M.A. With Maps. 6s.

CATULLUS.—SELECT POEMS. Edited by F. P. SIMPSON, B.A. 5s. The Text of this Edition is carefully expurgated for School use.

CICERO.—THE CATILINE ORATIONS. By A. S. WILKINS, Litt.D., Professor of Latin in the Owens College, Victoria University, Manchester. 3s. 6d.

PRO LEGE MANILIA. By Prof. A. S. WILKINS, Litt.D. 2s. 6d.

THE SECOND PHILIPPIC ORATION. By JOHN E. B. MAYOR, M.A., Professor of Latin in the University of Cambridge. 5s.

PRO ROSCIO AMERINO. By E. H. DONKIN, M.A. 4s. 6d.

PRO P. SESTIO. By Rev. H. A. HOLDEN, Litt.D. 5s.

DEMOSTHENES.—DE CORONA. By B. DRAKE, M.A. 7th Edition, revised by E. S. SHUCKBURGH, M.A. 4s. 6d.

ADVERSUS LEPTINEM. By Rev. J. R. KING, M.A., Fellow and Tutor of Oriel College, Oxford. 4s. 6d.

THE FIRST PHILIPPIC. By Rev. T. GWATKIN, M.A. 2s. 6d.

IN MIDIAM. By Prof. A. S. WILKINS, Litt.D., and HERMAN HAGER, Ph.D., of the Owens College, Victoria University, Manchester. *[In preparation.*

EURIPIDES.—HIPPOLYTUS. By Rev. J. P. MAHAFFY, D.D., Fellow of Trinity College, and Professor of Ancient History in the University of Dublin, and J. B. BURY, M.A., Fellow of Trinity College, Dublin. 3s. 6d.

MEDEA. By A. W. VERRALL, Litt.D., Fellow of Trinity College, Cambridge. 3s. 6d.

IPHIGENIA IN TAURIS. By E. B. ENGLAND, M.A. 4s. 6d.

ION. By M. A. BAYFIELD, M.A., Headmaster's Assistant at Malvern College. 3s. 6d.

BACCHAE. By R. Y. TYRRELL, M.A., Regius Professor of Greek in the University of Dublin. *[In preparation.*

HERODOTUS.—BOOK III. By G. C. MACAULAY, M.A. *[In the Press.*

BOOK V. By J. STRACHAN, M.A., Professor of Greek in the Owens College, Victoria University, Manchester. *[In preparation.*

BOOK VI. By the same. *[In the Press.*

BOOKS VII. and VIII. By Mrs. MONTAGU BUTLER. *[In the Press.*

HESIOD.—THE WORKS AND DAYS. By W. T. LENDRUM, M.A., Assistant Master at Dulwich College. *[In preparation.*

HOMER.—ILIAD. BOOKS I., IX., XI., XVI.—XXIV. THE STORY OF ACHILLES. By the late J. H. PRATT, M.A., and WALTER LEAF, Litt.D., Fellows of Trinity College, Cambridge. 6s.

ODYSSEY. BOOK IX. By Prof. JOHN E. B. MAYOR. 2s. 6d.

ODYSSEY. BOOKS XXI.—XXIV. THE TRIUMPH OF ODYSSEUS. By S. G. HAMILTON, B.A., Fellow of Hertford College, Oxford. 3s. 6d.

HORACE.—THE ODES. By T. E. PAGE, M.A., Assistant Master at the Charter-house. 6s. (BOOKS I., II., III., and IV. separately, 2s. each).

THE SATIRES. By ARTHUR PALMER, M.A., Professor of Latin in the University of Dublin. 6s.

THE EPISTLES AND ARS POETICA. By A. S. WILKINS, Litt.D., Professor of Latin in the Owens College, Victoria University, Manchester. 6s.

ISAEOS.—THE ORATIONS. By WILLIAM RIDGEWAY, M.A., Professor of Greek in Queen's College, Cork. [In preparation.

JUVENAL.—THIRTEEN SATIRES. By E. G. HARDY, M.A. 5s. The Text is carefully expurgated for School use.

SELECT SATIRES. By Prof. JOHN E. B. MAYOR. X. and XI. 3s. 6d. XII.—XVI. 4s. 6d.

LIVY. BOOKS II. and III. By Rev. H. M. STEPHENSON, M.A. 5s.

BOOKS XXI. and XXII. By Rev. W. W. CAPES, M.A. With Maps. 5s.

BOOKS XXIII. and XXIV. By G. C. MACAULAY, M.A. With Maps. 5s.

THE LAST TWO KINGS OF MACEDON. EXTRACTS FROM THE FOURTH AND FIFTH DECADES OF LIVY. By F. H. RAWLINS, M.A., Assistant Master at Eton. With Maps. 3s. 6d.

THE SUBJUGATION OF ITALY. SELECTIONS FROM THE FIRST DECADE. By G. E. MARINDIN, M.A. [In preparation.

LUCRETIUS.—BOOKS I.—III. By J. H. WARBURTON LEE, M.A., Assistant Master at Rossall. 4s. 6d.

LYSIAS.—SELECT ORATIONS. By E. S. SHUCKBURGH, M.A. 6s.

MARTIAL.—SELECT EPIGRAMS. By Rev. H. M. STEPHENSON, M.A. 6s. 6d.

OVID.—FASTI. By G. H. HALLAM, M.A., Assistant Master at Harrow. With Maps. 5s.

HEROIDUM EPISTULÆ XIII. By E. S. SHUCKBURGH, M.A. 4s. 6d.

METAMORPHOSES. BOOKS I.—III. By C. SIMMONS, M.A. [In preparation.

BOOKS XIII. and XIV. By the same Editor. 4s. 6d.

PLATO.—LACHES. By M. T. TATHAM, M.A. 2s. 6d.

THE REPUBLIC. BOOKS I.—V. By T. H. WARREN, M.A., President of Magdalen College, Oxford. 6s.

PLAUTUS.—MILES GLORIOSUS. By R. W. TYRRELL, M.A., Regius Professor of Greek in the University of Dublin. 2d Ed., revised. 5s.

AMPHITRUO. By ARTHUR PALMER, M.A., Professor of Latin in the University of Dublin. 5s.

PLINY.—LETTERS. BOOKS I. and II. By J. COWAN, M.A., Assistant Master at the Manchester Grammar School. 5s.

LETTERS. BOOK III. By Prof. JOHN E. B. MAYOR. With Life of Pliny by G. H. RENDALL, M.A. 5s.

PLUTARCH.—LIFE OF THEMISTOKLES. By Rev. H. A. HOLDEN, Litt.D. 5s.

LIVES OF GALBA AND OTHO. By E. G. HARDY, M.A. 6s.

POLYBIUS.—THE HISTORY OF THE ACHÆAN LEAGUE AS CONTAINED IN THE REMAINS OF POLYBIUS. By W. W. CAPES, M.A. 6s. 6d.

PROPERTIUS.—SELECT POEMS. By Prof. J. P. POSTGATE, Litt.D., Fellow of Trinity College, Cambridge. 2d Ed., revised. 6s.

SALLUST.—CATILINA and JUGURTHA. By C. MERIVALE, D.D., Dean of Ely. 4s. 6d. Or separately, 2s. 6d. each.

BELLUM CATULINÆ. By A. M. COOK, M.A., Assistant Master at St. Paul's School. 4s. 6d.

JUGURTHA. By the same Editor. [In preparation.

TACITUS.—THE ANNALS. BOOKS I. and II. By J. S. REID, Litt.D. [In preparation.

THE ANNALS. BOOK VI. By A. J. CHURCH, M.A., and W. J. BRODRIBB, M.A. 2s. 6d.

THE HISTORIES. BOOKS I. and II. By A. D. GODLEY, M.A., Fellow of Magdalen College, Oxford. 5s. BOOKS III.-V. By the same. 5s.
AGRICOLA and GERMANIA. By A. J. CHURCH, M.A., and W. J. BRODRIBB, M.A. 3s. 6d. Or separately, 2s. each.
TERENCE.—HAUTON TIMORUMENOS. By E. S. SHUCKBURGH, M.A. 3s. With Translation. 4s. 6d.
PHORMIO. By Rev. JOHN BOND, M.A., and Rev. A S. WALPOLE, M.A. 4s. 6d.
THUCYDIDES.—BOOK I. By C. BRYANS, M.A. [In preparation.
BOOK II. By E. C. MARCHANT, M.A., Assistant Master at St. Paul's School.
                                                                [In preparation.
BOOK III. By C. BRYANS. [In preparation.
BOOK IV. By C. E. GRAVES, M.A., Classical Lecturer at St. John's College, Cambridge. 5s.
BOOK V. By the same Editor. [In the Press.
BOOKS VI. AND VII. THE SICILIAN EXPEDITION. By Rev. PERCIVAL FROST, M.A. With Map. 5s.
BOOK VIII. By Prof. T. G. TUCKER, M.A. [In preparation.
TIBULLUS.—SELECT POEMS. By Prof. J. P. POSTGATE, Litt.D. [In preparation.
VIRGIL.—ÆNEID. BOOKS II. AND III. THE NARRATIVE OF ÆNEAS. By E. W. HOWSON, M.A., Assistant Master at Harrow. 3s.
XENOPHON.—THE ANABASIS. BOOKS I.-IV. By Profs. W. W. GOODWIN and J. W. WHITE. Adapted to Goodwin's Greek Grammar. With Map. 5s.
HELLENICA. BOOKS I. AND II. By H. HAILSTONE, B.A. With Map. 4s. 6d.
CYROPÆDIA. BOOKS VII. AND VIII. By A. GOODWIN, M.A., Professor of Classics in University College, London. 5s.
MEMORABILIA SOCRATIS. By A. R. CLUER, B.A., Balliol College, Oxford. 6s.
HIERO. By Rev H. A. HOLDEN, Litt.D., LL.D. 3s. 6d.
OECONOMICUS. By the same. With Lexicon. 6s.

CLASSICAL LIBRARY.

Texts, Edited with **Introductions and Notes,** for the use of Advanced Students ; **Commentaries and Translations.**

ÆSCHYLUS.—THE SUPPLICES. A Revised Text, with Translation. By T. G. TUCKER, M.A., Professor of Classical Philology in the University of Melbourne. 8vo. 10s. 6d.
THE SEVEN AGAINST THEBES. With Translation. By A. W. VERRALL, Litt.D., Fellow of Trinity College, Cambridge. 8vo. 7s. 6d.
AGAMEMNON. With Translation. By A W. VERRALL, Litt.D. 8vo. 12s.
AGAMEMNON, CHOEPHORŒ, AND EUMENIDES. By A. O. PRICKARD, M.A., Fellow and Tutor of New College, Oxford. 8vo. [In preparation.
THE EUMENIDES. With Verse Translation. By BERNARD DRAKE, M.A. 8vo. 5s.
ANTONINUS, MARCUS AURELIUS.—BOOK IV. OF THE MEDITATIONS. With Translation. By HASTINGS CROSSLEY, M.A. 8vo. 6s.
ARISTOTLE.—THE METAPHYSICS. BOOK I. Translated by a Cambridge Graduate. 8vo. 5s.
THE POLITICS. By R. D. HICKS, M.A., Fellow of Trinity College, Cambridge. 8vo. [In the Press.
THE POLITICS. Translated by Rev. J. E. C. WELLDON, M.A., Headmaster of Harrow. Cr. 8vo. 10s. 6d.
THE RHETORIC. Translated by the same. Cr. 8vo. 7s. 6d.
AN INTRODUCTION TO ARISTOTLE'S RHETORIC. With Analysis, Notes, and Appendices. By E. M. COPE, Fellow and late Tutor of Trinity College, Cambridge. 8vo. 14s.

THE ETHICS. Translated by Rev. J. E. C. WELLDON, M.A. Cr. 8vo.
[*In preparation.*
THE SOPHISTICI ELENCHI. With Translation. By E. POSTE, M.A., Fellow of Oriel College, Oxford. 8vo. 8s. 6d.
ARISTOPHANES.—THE BIRDS. Translated into English Verse. By B. H. KENNEDY, D.D. Cr. 8vo. 6s. Help Notes to the Same, for the Use of Students. 1s. 6d.
ATTIC ORATORS.—FROM ANTIPHON TO ISAEOS. By R. C. JEBB, Litt.D., Regius Professor of Greek in the University of Cambridge. 2 vols. 8vo. 25s.
BABRIUS.—With Lexicon. By Rev. W. G. RUTHERFORD, M.A., LL.D., Head-master of Westminster. 8vo. 12s. 6d.
CICERO.—THE ACADEMICA. By J. S. REID, Litt.D., Fellow of Caius College, Cambridge. 8vo. 15s.
THE ACADEMICS. Translated by the same. 8vo. 5s. 6d.
SELECT LETTERS. After the Edition of ALBERT WATSON, M.A. Translated by G. E. JEANS, M.A., Fellow of Hertford College, Oxford. Cr. 8vo. 10s. 6d.
EURIPIDES.—MEDEA. Edited by A. W. VERRALL, Litt.D. 8vo. 7s. 6d.
IPHIGENIA IN AULIS. Edited by E. B. ENGLAND, M.A. 8vo. [*In the Press.*
INTRODUCTION TO THE STUDY OF EURIPIDES. By Professor J. P. MAHAFFY. Fcap. 8vo. 1s. 6d. (*Classical Writers*).
HERODOTUS.—BOOKS I.-III. THE ANCIENT EMPIRES OF THE EAST. Edited by A. H. SAYCE, Deputy-Professor of Comparative Philology, Oxford. 8vo. 16s.
BOOKS IV.-IX. Edited by R. W. MACAN, M.A., Lecturer in Ancient History at Brasenose College, Oxford. 8vo. [*In preparation.*
THE HISTORY. Translated by G. C. MACAULAY, M.A. 2 vols. Cr. 8vo. 18s.
HOMER.—THE ILIAD. By WALTER LEAF, Litt.D. 8vo. Books I.-XII. 14s. Books XIII.-XXIV. 14s.
THE ILIAD. Translated into English Prose by ANDREW LANG, M.A., WALTER LEAF, Litt.D., and ERNEST MYERS, M.A. Cr. 8vo. 12s. 6d.
THE ODYSSEY. Done into English by S. H. BUTCHER, M.A., Professor of Greek in the University of Edinburgh, and ANDREW LANG, M.A. Cr. 8vo. 6s.
INTRODUCTION TO THE STUDY OF HOMER. By the Right Hon. W. E. GLADSTONE. 18mo. 1s. (*Literature Primers*.)
HOMERIC DICTIONARY. Translated from the German of Dr. G. AUTENRIETH by R. P. KEEP, Ph.D. Illustrated. Cr. 8vo. 6s.
HORACE.—Translated by J. LONSDALE, M.A., and S. LEE, M.A. Gl. 8vo. 3s. 6d.
STUDIES, LITERARY AND HISTORICAL, IN THE ODES OF HORACE. By A. W. VERRALL, Litt.D. 8vo. 8s. 6d.
JUVENAL.—THIRTEEN SATIRES OF JUVENAL. By JOHN E. B. MAYOR, M.A., Professor of Latin in the University of Cambridge. Cr. 8vo. 2 vols. 10s. 6d. each. Vol. I. 10s. 6d. Vol. II. 10s. 6d.
THIRTEEN SATIRES. Translated by ALEX. LEEPER, M.A., LL.D., Warden of Trinity College, Melbourne. Cr. 8vo. 3s. 6d.
KTESIAS.—THE FRAGMENTS OF THE PERSIKA OF KTESIAS. By JOHN GILMORE, M.A. 8vo. 8s. 6d.
LIVY.—BOOKS I.-IV. Translated by Rev. H. M. STEPHENSON, M.A.
[*In preparation.*
BOOKS XXI.-XXV. Translated by A. J. CHURCH, M.A., and W. J. BRODRIBB, M.A. Cr. 8vo. 7s. 6d.
INTRODUCTION TO THE STUDY OF LIVY. By Rev. W. W. CAPES, M.A. Fcap. 8vo. 1s. 6d. (*Classical Writers*.)
MARTIAL.—BOOKS I. AND II. OF THE EPIGRAMS. By Prof. JOHN E. B. MAYOR, M.A. 8vo. [*In the Press.*
PAUSANIAS.—DESCRIPTION OF GREECE. Translated with Commentary by J. G. FRAZER, M.A., Fellow of Trinity College, Cambridge.
[*In preparation.*

PHRYNICHUS.—THE NEW PHRYNICHUS ; being a Revised Text of the Ecloga of the Grammarian Phrynichus. With Introduction and Commentary by Rev. W. G. RUTHERFORD, M.A., LL.D., Headmaster of Westminster. 8vo. 18s.

PINDAR.—THE EXTANT ODES OF PINDAR. Translated by ERNEST MYERS, M.A. Cr. 8vo. 5s.

THE OLYMPIAN AND PYTHIAN ODES. Edited, with an Introductory Essay, by BASIL GILDERSLEEVE, Professor of Greek in the Johns Hopkins University, U.S.A. Cr. 8vo. 7s. 6d.

THE NEMEAN ODES. By J. B. BURY, M.A., Fellow of Trinity College, Dublin. 8vo. *[In the Press.*

PLATO.—PHÆDO. By R. D. ARCHER-HIND, M.A., Fellow of Trinity College, Cambridge. 8vo. 8s. 6d.

PHÆDO. By W. D. GEDDES, LL.D., Principal of the University of Aberdeen. 8vo. 8s. 6d.

TIMAEUS. With Translation. By R. D. ARCHER-HIND, M.A. 8vo. 16s.

THE REPUBLIC OF PLATO. Translated by J. LL. DAVIES, M.A., and D. J VAUGHAN, M.A. 18mo. 4s. 6d.

EUTHYPHRO, APOLOGY, CRITO, AND PHÆDO. Translated by F. J. CHURCH. 18mo. 4s. 6d.

PHÆDRUS, LYSIS, AND PROTAGORAS. Translated by J. WRIGHT, M.A. 18mo. 4s. 6d.

PLAUTUS.—THE MOSTELLARIA. By WILLIAM RAMSAY, M.A. Edited by G. G. RAMSAY, M.A., Professor of Humanity in the University of Glasgow 8vo. 14s.

PLINY.—CORRESPONDENCE WITH TRAJAN. C. Plinii Caecilii Secundi Epistulae ad Traianum Imperatorem cum Eiusdem Responsis. By E. G. HARDY, M.A. 8vo. 10s. 6d.

POLYBIUS.—THE HISTORIES OF POLYBIUS. Translated by E. S. SHUCK-BURGH, M.A. 2 vols. Cr. 8vo. 24s.

SALLUST.—CATILINE AND JUGURTHA. Translated by A. W. POLLARD, B.A. Cr. 8vo. 6s. THE CATILINE (separately). 3s.

SOPHOCLES—ŒDIPUS THE KING. Translated into English Verse by E. D. A. MORSHEAD, M.A., Assistant Master at Winchester. Fcap. 8vo. 3s. 6d.

TACITUS.—THE ANNALS. By G. O. HOLBROOKE, M.A., Professor of Latin in Trinity College, Hartford, U.S.A. With Maps. 8vo. 16s.

THE ANNALS. Translated by A. J. CHURCH, M.A., and W. J. BRODRIBB, M.A. With Maps. Cr. 8vo. 7s. 6d.

THE HISTORIES. By Rev. W. A. SPOONER, M.A., Fellow of New College, Oxford. 8vo. *[In the Press.*

THE HISTORY. Translated by A. J. CHURCH, M.A., and W. J. BRODRIBB, M.A. With Map. Cr. 8vo. 6s.

THE AGRICOLA AND GERMANY, WITH THE DIALOGUE ON ORATORY. Translated by A. J. CHURCH, M.A., and W. J. BRODRIBB, M.A. With Maps. Cr. 8vo. 4s. 6d.

INTRODUCTION TO THE STUDY OF TACITUS. By A. J. CHURCH, M.A., and W. J. BRODRIBB, M.A. Fcap. 8vo. 1s. 6d. (*Classical Writers.*)

THEOCRITUS, BION, AND MOSCHUS. Translated by A. LANG, M.A. 18mo 4s. 6d.

*** Also an Edition on Large Paper. Cr. 8vo. 9s.

THUCYDIDES.—BOOK IV. A Revision of the Text, Illustrating the Principal Causes of Corruption in the Manuscripts of this Author. By Rev. W. G. RUTHERFORD, M.A., LL.D., Headmaster of Westminster.

BOOK VIII. By H. C. GOODHART, M.A., Fellow of Trinity College, Cambridge. *[In the Press.*

VIRGIL.—Translated by J. LONSDALE, M.A., and S. LEE, M.A. Gl. 8vo. 3s. 6d.

THE ÆNEID. Translated by J. W. MACKAIL, M.A., Fellow of Balliol College, Oxford. Cr. 8vo. 7s. 6d.

XENOPHON.—Translated by H. G. DAKYNS, M.A. In four vols. Vol. I., containing "The Anabasis" and Books I. and II. of "The Hellenica." Cr. 8vo 10s. 6d. *[Vol. II. in the Press*

GRAMMAR, COMPOSITION, & PHILOLOGY.

BELCHER.—SHORT EXERCISES IN LATIN PROSE COMPOSITION AND EXAMINATION PAPERS IN LATIN GRAMMAR. Part I. By Rev. H. BELCHER, LL.D., Rector of the High School, Dunedin, N.Z. 18mo. 1s. 6d. KEY, for Teachers only. 18mo. 3s. 6d.

Part II., On the Syntax of Sentences, with an Appendix, including EXERCISES IN LATIN IDIOMS, etc. 18mo. 2s. KEY, for Teachers only. 18mo. 3s.

BLACKIE.—GREEK AND ENGLISH DIALOGUES FOR USE IN SCHOOLS AND COLLEGES. By JOHN STUART BLACKIE, Emeritus Professor of Greek in the University of Edinburgh. New Edition. Fcap. 8vo. 2s. 6d.

BRYANS.—LATIN PROSE EXERCISES BASED UPON CAESAR'S GALLIC WAR. With a Classification of Cæsar's Chief Phrases and Grammatical Notes on Cæsar's Usages. By CLEMENT BRYANS, M.A., Assistant Master at Dulwich College. Ex. fcap. 8vo. 2s. 6d. KEY, for Teachers only. 4s. 6d.

GREEK PROSE EXERCISES based upon Thucydides. By the same.
[*In preparation.*]

COOKSON.—A LATIN SYNTAX. By CHRISTOPHER COOKSON, M.A., Assistant Master at St. Paul's School. 8vo. [*In preparation.*]

CORNELL UNIVERSITY STUDIES IN CLASSICAL PHILOLOGY. Edited by I. FLAGG, W. G. HALE, and B. I. WHEELER. I. The *CUM*-Constructions: their History and Functions. By W. G. HALE. Part 1. Critical. 1s. 8d. nett. Part 2. Constructive. 3s. 4d. nett. II. Analogy and the Scope of its Application in Language. By B. I. WHEELER. 1s. 3d. nett.

EICKE.—FIRST LESSONS IN LATIN. By K. M. EICKE, B.A., Assistant Master at Oundle School. Gl. 8vo. 2s. 6d.

ENGLAND.—EXERCISES ON LATIN SYNTAX AND IDIOM. ARRANGED WITH REFERENCE TO ROBY'S SCHOOL LATIN GRAMMAR. By E. B. ENGLAND, Assistant Lecturer at the Owens College, Victoria University, Manchester. Cr. 8vo. 2s. 6d. KEY, for Teachers only. 2s. 6d.

GILES.—A MANUAL OF GREEK AND LATIN PHILOLOGY. By P. GILES, M.A., Fellow of Gonville and Caius College, Cambridge. Cr. 8vo.
[*In the Press.*]

GOODWIN.—Works by W. W. GOODWIN, LL.D., D.C.L., Professor of Greek in Harvard University, U.S.A.

SYNTAX OF THE MOODS AND TENSES OF THE GREEK VERB. New Ed., revised and enlarged. 8vo. 14s.

A GREEK GRAMMAR. Cr. 8vo. 6s.

A GREEK GRAMMAR FOR SCHOOLS. Cr. 8vo. 3s. 6d.

GREENWOOD.—THE ELEMENTS OF GREEK GRAMMAR. Adapted to the System of Crude Forms. By J. G. GREENWOOD, sometime Principal of the Owens College, Manchester. Cr. 8vo. 5s. 6d.

HADLEY AND ALLEN.—A GREEK GRAMMAR FOR SCHOOLS AND COLLEGES. By JAMES HADLEY, late Professor in Yale College. Revised and in part rewritten by F. DE F. ALLEN, Professor in Harvard College. Cr. 8vo. 6s.

HODGSON.—MYTHOLOGY FOR LATIN VERSIFICATION. A brief sketch of the Fables of the Ancients, prepared to be rendered into Latin Verse for Schools. By F. HODGSON, B.D., late Provost of Eton. New Ed., revised by F. C. HODGSON, M.A. 18mo. 3s.

JACKSON.—FIRST STEPS TO GREEK PROSE COMPOSITION. By BLOMFIELD JACKSON, M.A., Assistant Master at King's College School. 18mo. 1s. 6d. KEY, for Teachers only. 18mo. 3s. 6d.

SECOND STEPS TO GREEK PROSE COMPOSITION, with Miscellaneous Idioms, Aids to Accentuation, and Examination Papers in Greek Scholarship. By the same. 18mo. 2s. 6d. KEY, for Teachers only. 18mo. 3s. 6d.

KYNASTON.—EXERCISES IN THE COMPOSITION OF GREEK IAMBIC VERSE by Translations from English Dramatists. By Rev. H. KYNASTON, D.D., Professor of Classics in the University of Durham. With Vocabulary Ex. fcap. 8vo. 5s.

KEY, for Teachers only. Ex. fcap. 8vo. 4s. 6d.

LUPTON.- AN INTRODUCTION TO LATIN ELEGIAC VERSE COMPOSITION. By J. H. LUPTON, Sur-Master of St. Paul's School. Gl. 8vo. 2s. 6d.
KEY TO PART II. (XXV.—C.) Gl. 8vo. 3s. 6d.
AN INTRODUCTION TO LATIN LYRIC VERSE COMPOSITION. By the same. Gl. 8vo. 3s. KEY, for Teachers only. Gl. 8vo. 4s. 6d.

MACKIE.—PARALLEL PASSAGES FOR TRANSLATION INTO GREEK AND ENGLISH. With Indexes. By Rev. ELLIS C. MACKIE, M.A., Classical Master at Heversham Grammar School. Gl. 8vo. 4s. 6d.

MACMILLAN.—FIRST LATIN GRAMMAR. By M. C. MACMILLAN, M.A. Fcap. 8vo. 1s. 6d.

MACMILLAN'S GREEK COURSE.—Edited by Rev. W. G. RUTHERFORD, M.A., LL.D., Headmaster of Westminster. Gl. 8vo.
FIRST GREEK GRAMMAR— ACCIDENCE. By the Editor. 2s.
FIRST GREEK GRAMMAR—SYNTAX. By the same. 2s.
ACCIDENCE AND SYNTAX. In one volume. 3s. 6d.
EASY EXERCISES IN GREEK ACCIDENCE. By H. G. UNDERHILL, M.A., Assistant Master at St. Paul's Preparatory School. 2s.
A SECOND GREEK EXERCISE BOOK. By Rev. W. A. HEARD, M.A., Headmaster of Fettes College, Edinburgh. 2s. 6d.
MANUAL OF GREEK ACCIDENCE. By the Editor.   [In preparation.
MANUAL OF GREEK SYNTAX. By the Editor.   [In preparation.
ELEMENTARY GREEK COMPOSITION. By the Editor.   [In preparation.

MACMILLAN'S GREEK READER.—STORIES AND LEGENDS. A First Greek Reader, with Notes, Vocabulary, and Exercises. By F. H. COLSON, M.A., Headmaster of Plymouth College. Gl. 8vo. 3s.

MACMILLAN'S LATIN COURSE.—By A. M. COOK, M.A., Assistant Master at St. Paul's School. First Part. Gl. 8vo. 3s. 6d. Second Part. 2s. 6d.
          [Third Part in preparation.

MACMILLAN'S SHORTER LATIN COURSE.—By A. M. COOK, M.A. Being an abridgment of "Macmillan's Latin Course," First Part. Gl. 8vo. 1s. 6d.

MACMILLAN'S LATIN READER.—A LATIN READER FOR THE LOWER FORMS IN SCHOOLS. By H. J. HARDY, M.A., Assistant Master at Winchester. Gl. 8vo. 2s. 6d.

MARSHALL.—A TABLE OF IRREGULAR GREEK VERBS, classified according to the arrangement of Curtius's Greek Grammar. By J. M. MARSHALL, M.A., Headmaster of the Grammar School, Durham. 8vo. 1s.

MAYOR.—FIRST GREEK READER. By Prof. JOHN E. B. MAYOR, M.A., Fellow of St. John's College, Cambridge. Fcap. 8vo. 4s. 6d.

MAYOR.—GREEK FOR BEGINNERS.—By Rev. J. B. MAYOR, M.A., late Professor of Classical Literature in King's College, London. Part I., with Vocabulary, 1s. 6d. Parts II. and III., with Vocabulary and Index. Fcap. 8vo. 3s. 6d. Complete in one Vol. 4s. 6d.

NIXON.—PARALLEL EXTRACTS, Arranged for Translation into English and Latin, with Notes on Idioms. By J. E. NIXON, M.A., Fellow and Classical Lecturer, King's College, Cambridge. Part I.—Historical and Epistolary. Cr. 8vo. 3s. 6d.
PROSE EXTRACTS, Arranged for Translation into English and Latin, with General and Special Prefaces on Style and Idiom. By the same. I. Oratorical. II. Historical. III. Philosophical. IV. Anecdotes and Letters. 2d Ed., enlarged to 280 pp. Cr. 8vo. 4s. 6d.
SELECTIONS FROM PROSE EXTRACTS, including Easy Anecdotes and Letters and Notes and Hints. By the same. 120 pp. 3s.
          Translations of about 70 Extracts can be supplied to Schoolmasters (2s. 6d.), on application to the Author : and about 40 similarly of "Parallel Extracts," 1s. 6d. post free.

PANTIN.—A FIRST LATIN VERSE BOOK. By W. E. P. PANTIN, M.A., Assistant Master at St. Paul's School. Gl. 8vo. 1s. 6d.

PEILE.—A PRIMER OF PHILOLOGY. By J. PEILE, Litt. D., Master of Christ's College, Cambridge. 18mo. 1s.

POSTGATE.—SERMO LATINUS. A short Guide to Latin Prose Composition. By Prof. J. P. POSTGATE, Litt.D., Fellow of Trinity College, Cambridge. Gl. 8vo. 2s. 6d. KEY to "Selected Passages." Gl. 8vo. 3s. 6d.

POSTGATE AND VINCE.—A DICTIONARY OF LATIN ETYMOLOGY. By J. P. POSTGATE and C. A. VINCE. *[In preparation.*

POTTS.—HINTS TOWARDS LATIN PROSE COMPOSITION. By A. W. POTTS, M.A., LL.D., late Fellow of St. John's College, Cambridge. Ex. fcap. 8vo. 3s.

PASSAGES FOR TRANSLATION INTO LATIN PROSE. Edited with Notes and References to the above. Ex. fcap. 8vo. 2s. 6d. KEY, for Teachers only. 2s. 6d.

PRESTON.—EXERCISES IN LATIN VERSE OF VARIOUS KINDS. By Rev. G. PRESTON. Gl. 8vo. 2s. 6d. KEY, for Teachers only. Gl. 8vo. 5s.

REID.—A GRAMMAR OF TACITUS. By J. S. REID, Litt. D., Fellow of Caius College, Cambridge. *[In the Press.*

A GRAMMAR OF VIRGIL. By the same. *[In preparation.*

ROBY.—Works by H. J. ROBY, M.A., late Fellow of St. John's College, Cambridge.

A GRAMMAR OF THE LATIN LANGUAGE, from Plautus to Suetonius. Part I. Sounds, Inflexions, Word-formation, Appendices. Cr. 8vo. 9s. Part II. Syntax, Prepositions, etc. 10s. 6d.

SCHOOL LATIN GRAMMAR. Cr. 8vo. 5s.

RUSH.—SYNTHETIC LATIN DELECTUS. With Notes and Vocabulary. By E. RUSH, B.A. Ex. fcap. 8vo. 2s. 6d.

RUST.—FIRST STEPS TO LATIN PROSE COMPOSITION. By Rev. G. RUST, M.A. 18mo. 1s. 6d. KEY, for Teachers only. By W. M. YATES. 18mo. 3s. 6d.

RUTHERFORD.—Works by the Rev. W. G. RUTHERFORD, M.A., LL.D., Head-master of Westminster.

REX LEX. A Short Digest of the principal Relations between the Latin, Greek, and Anglo-Saxon Sounds. 8vo. *[In preparation.*

THE NEW PHRYNICHUS; being a Revised Text of the Ecloga of the Grammarian Phrynichus. With Introduction and Commentary. 8vo. 18s. (See also *Macmillan's Greek Course.*)

SHUCKBURGH.—PASSAGES FROM LATIN AUTHORS FOR TRANSLATION INTO ENGLISH. Selected with a view to the needs of Candidates for the Cambridge Local, and Public Schools' Examinations. By E. S. SHUCKBURGH, M.A. Cr. 8vo. 2s.

SIMPSON. — LATIN PROSE AFTER THE BEST AUTHORS: Cæsarian Prose. By F. P. SIMPSON, B.A. Ex. fcap. 8vo. 2s. 6d. KEY, for Teachers only. Ex. fcap. 8vo. 5s.

STRACHAN AND WILKINS.—ANALECTA. Selected Passages for Translation. By J. S. STRACHAN, M.A., Professor of Greek, and A. S. WILKINS, Litt. D., Professor of Latin in the Owens College, Manchester. Cr. 8vo. 5s.

THRING.—Works by the Rev. E. THRING, M.A., late Headmaster of Uppingham.

A LATIN GRADUAL. A First Latin Construing Book for Beginners. With Coloured Sentence Maps. Fcap. 8vo. 2s. 6d.

A MANUAL OF MOOD CONSTRUCTIONS. Fcap. 8vo. 1s. 6d.

WELCH AND DUFFIELD. — LATIN ACCIDENCE AND EXERCISES ARRANGED FOR BEGINNERS. By W. WELCH and C. G. DUFFIELD, Assistant Masters at Cranleigh School. 18mo. 1s. 6d.

WHITE.—FIRST LESSONS IN GREEK. Adapted to GOODWIN'S GREEK GRAMMAR, and designed as an introduction to the ANABASIS OF XENOPHON. By JOHN WILLIAMS WHITE, Assistant-Professor of Greek in Harvard University, U.S.A. Cr. 8vo. 4s. 6d.

WRIGHT.—Works by J. WRIGHT, M.A., late Headmaster of Sutton Coldfield School.

A HELP TO LATIN GRAMMAR; or, the Form and Use of Words in Latin, with Progressive Exercises. Cr. 8vo. 4s. 6d.

THE SEVEN KINGS OF ROME. An Easy Narrative, abridged from the First Book of Livy by the omission of Difficult Passages; being a First Latin Reading Book, with Grammatical Notes and Vocabulary. Fcap. 8vo. 3s. 6d.

FIRST LATIN STEPS; OR, AN INTRODUCTION BY A SERIES OF EXAMPLES TO THE STUDY OF THE LATIN LANGUAGE. Cr. 8vo. 3s.

ATTIC PRIMER. Arranged for the Use of Beginners. Ex. fcap. 8vo. 2s. 6d.

A COMPLETE LATIN COURSE, comprising Rules with Examples, Exercises, both Latin and English, on each Rule, and Vocabularies. Cr. 8vo. 2s. 6d.

ANTIQUITIES, ANCIENT HISTORY, AND PHILOSOPHY.

ARNOLD.—A HANDBOOK OF LATIN EPIGRAPHY. By W. T. ARNOLD, M.A. [In preparation.

THE ROMAN SYSTEM OF PROVINCIAL ADMINISTRATION TO THE ACCESSION OF CONSTANTINE THE GREAT. By the same. Cr. 8vo. 6s.

ARNOLD.—THE SECOND PUNIC WAR. Being Chapters from THE HISTORY OF ROME by the late THOMAS ARNOLD, D.D., Headmaster of Rugby. Edited, with Notes, by W. T. ARNOLD, M.A. With 8 Maps. Cr. 8vo. 8s. 6d.

BEESLY.—STORIES FROM THE HISTORY OF ROME. By Mrs. BEESLY. Fcap. 8vo. 2s. 6d.

BURN.—ROMAN LITERATURE IN RELATION TO ROMAN ART. By Rev. ROBERT BURN, M.A., late Fellow of Trinity College, Cambridge. Illustrated. Ex. cr. 8vo. 14s.

BURY.—A HISTORY OF THE LATER ROMAN EMPIRE FROM ARCADIUS TO IRENE, A.D. 395-800. By J. B. BURY, M.A., Fellow of Trinity College, Dublin. 2 vols. 8vo. 32s.

CLASSICAL WRITERS.—Edited by JOHN RICHARD GREEN, M.A., LL.D. Fcap. 8vo. 1s. 6d. each.

SOPHOCLES. By Prof. L. CAMPBELL, M.A.

EURIPIDES. By Prof. MAHAFFY, D.D.

DEMOSTHENES. By Prof. S. H. BUTCHER, M.A.

VIRGIL. By Prof. NETTLESHIP, M.A.

LIVY. By Rev. W. W. CAPES, M.A.

TACITUS. By Prof. A. J. CHURCH, M.A., and W. J. BRODRIBB, M.A.

MILTON. By Rev. STOPFORD A. BROOKE, M.A.

FREEMAN.—Works by EDWARD A. FREEMAN, D.C.L., LL.D., Regius Professor of Modern History in the University of Oxford.

HISTORY OF ROME. (Historical Course for Schools.) 18mo. [In preparation.

HISTORY OF GREECE. (Historical Course for Schools.) 18mo. [In preparation.

A SCHOOL HISTORY OF ROME. Cr. 8vo. [In preparation.

HISTORICAL ESSAYS. Second Series. [Greek and Roman History.] 8vo. 10s. 6d.

FYFFE.—A SCHOOL HISTORY OF GREECE. By C. A. FYFFE, M.A. Cr. 8vo. [In preparation.

GARDNER.—SAMOS AND SAMIAN COINS. An Essay. By PERCY GARDNER, Litt.D., Professor of Archæology in the University of Oxford. With Illustrations. 8vo. 7s. 6d.

GEDDES.—THE PROBLEM OF THE HOMERIC POEMS. By W. D. GEDDES Principal of the University of Aberdeen. 8vo. 14s.

GLADSTONE.—Works by the Rt. Hon. W. E. GLADSTONE, M.P.

THE TIME AND PLACE OF HOMER. Cr. 8vo. 6s. 6d.

A PRIMER OF HOMER. 18mo. 1s.

GOW.—A COMPANION TO SCHOOL CLASSICS. By JAMES GOW, Litt.D., Master of the High School, Nottingham. With Illustrations. 2d Ed., revised. Cr. 8vo. 6s.

HARRISON AND VERRALL.—MYTHOLOGY AND MONUMENTS OF ANCIENT ATHENS. Translation of a portion of the "Attica" of Pausanias. By MARGARET DE G. VERRALL. With Introductory Essay and Archæological Commentary by JANE E. HARRISON. With Illustrations and Plans. Cr. 8vo. 16s.

JEBB.—Works by R. C. JEBB, Litt.D., Professor of Greek in the University of Cambridge.

THE ATTIC ORATORS FROM ANTIPHON TO ISAEOS. 2 vols. 8vo. 25s.

A PRIMER OF GREEK LITERATURE. 18mo. 1s.
(See also *Classical Series*.)

KIEPERT. — MANUAL OF ANCIENT GEOGRAPHY. By Dr. H KIE-PERT. Cr. 8vo. 5s.

LANCIANI.—ANCIENT ROME IN THE LIGHT OF RECENT DISCOVERIES.—By RODOLFO LANCIANI, Professor of Archæology in the University of Rome. Illustrated. 4to. 24s.

MAHAFFY.—Works by J. P. MAHAFFY, D.D., Fellow of Trinity College, Dublin, and Professor of Ancient History in the University of Dublin.

SOCIAL LIFE IN GREECE; from Homer to Menander. Cr. 8vo. 9s.

GREEK LIFE AND THOUGHT; from the Age of Alexander to the Roman Conquest. Cr. 8vo. 12s. 6d.

THE GREEK WORLD UNDER ROMAN SWAY. From Plutarch to Polybius. Cr. 8vo. [*In the Press.*

RAMBLES AND STUDIES IN GREECE. With Illustrations. With Map. Cr. 8vo. 10s. 6d.

A HISTORY OF CLASSICAL GREEK LITERATURE. In 2 vols. Cr. 8vo. Vol. I. The Poets, with an Appendix on Homer by Prof. SAYCE. 9s. Vol. II. The Prose Writers. In two parts.

A PRIMER OF GREEK ANTIQUITIES. With Illustrations. 18mo. 1s.

EURIPIDES. 18mo. 1s. 6d. (*Classical Writers*.)

MAYOR.—BIBLIOGRAPHICAL CLUE TO LATIN LITERATURE. Edited after HÜBNER. With large Additions. By Prof. JOHN E. B. MAYOR. Cr. 8vo. 10s. 6d.

NEWTON.—ESSAYS IN ART AND ARCHÆOLOGY. By Sir CHARLES NEWTON, K.C.B., D.C.L. 8vo. 12s. 6d.

SAYCE.—THE ANCIENT EMPIRES OF THE EAST. By A. H. SAYCE, M.A., Deputy-Professor of Comparative Philology, Oxford. Cr. 8vo. 6s.

SHUCKBURGH.—A SCHOOL HISTORY OF ROME. By E. S. SHUCKBURGH, M.A. Cr. 8vo. [*In preparation.*

STEWART.—THE TALE OF TROY. Done into English by AUBREY STEWART. Gl. 8vo. 3s. 6d.

WALDSTEIN.—CATALOGUE OF CASTS IN THE MUSEUM OF CLASSICAL ARCHÆOLOGY, CAMBRIDGE. By CHARLES WALDSTEIN, University Reader in Classical Archæology. Cr. 8vo. 1s. 6d.
** Also an Edition on Large Paper, small 4to. 5s.

WILKINS.—Works by Prof. WILKINS, Litt.D., LL.D.

A PRIMER OF ROMAN ANTIQUITIES. Illustrated. 18mo. 1s.

A PRIMER OF ROMAN LITERATURE. 18mo. 1s.

WILKINS AND ARNOLD.—A MANUAL OF ROMAN ANTIQUITIES. By Prof. A. S. WILKINS, Litt.D., and W. T. ARNOLD, M.A. Cr. 8vo. Illustrated. [*In preparation.*

MODERN LANGUAGES AND LITERATURE.

English; French; German; Modern Greek; Italian; Spanish.

ENGLISH.

ABBOTT.—A SHAKESPEARIAN GRAMMAR An Attempt to Illustrate some of the Differences between Elizabethan and Modern English. By the Rev. E. A. ABBOTT, D.D., formerly Headmaster of the City of London School. Ex. fcap. 8vo. 6s.

BACON.—ESSAYS. With Introduction and Notes, by F. G. SELBY, M.A., Professor of Logic and Moral Philosophy, Deccan College, Poona. Gl. 8vo. 3s. 6d.

BURKE.—REFLECTIONS ON THE FRENCH REVOLUTION. By the same. Gl. 8vo. [In July.

BROOKE.—PRIMER OF ENGLISH LITERATURE. By Rev. STOPFORD A. BROOKE, M.A. 18mo. 1s.

EARLY ENGLISH LITERATURE. By the same. 2 vols. 8vo. [In preparation.

BUTLER.—HUDIBRAS. With Introduction and Notes, by ALFRED MILNES, M.A. Ex. fcap. 8vo. Part I. 3s. 6d. Parts II. and III. 4s. 6d.

CAMPBELL.—SELECTIONS. With Introduction and Notes, by CECIL M. BARROW, M.A., Principal and Professor of English and Classics, Doveton College, Madras. Gl. 8vo. [In preparation.

COWPER.—THE TASK: an Epistle to Joseph Hill, Esq.; TIROCINIUM, or a Review of the Schools; and THE HISTORY OF JOHN GILPIN. Edited, with Notes, by W. BENHAM, B.D. Gl. 8vo. 1s. (Globe Readings from Standard Authors.)

THE TASK. With Introduction and Notes, by F. J. ROWE, M.A., and W. T. WEBB, M.A., Professors of English Literature, Presidency College, Calcutta. [In preparation.

DOWDEN.—SHAKESPERE. By Prof. DOWDEN. 18mo. 1s.

DRYDEN.—SELECT PROSE WORKS. Edited, with Introduction and Notes, by Prof. C. D. YONGE. Fcap. 8vo. 2s. 6d.

GLOBE READERS. For Standards I.-VI. Edited by A. F. MURISON. Illustrated. Gl. 8vo.

Primer I. (48 pp.)	3d.	Book III. (232 pp.)	1s. 3d.
Primer II. (48 pp.)	3d.	Book IV. (328 pp.)	1s. 9d.
Book I. (96 pp.)	6d.	Book V. (416 pp.)	2s.
Book II. (136 pp.)	9d.	Book VI. (448 pp.)	2s. 6d.

*THE SHORTER GLOBE READERS.**—Illustrated. Gl. 8vo.

Primer I. (48 pp.)	3d.	Standard III. (178 pp.) 1s.
Primer II. (48 pp.)	3d.	Standard IV. (182 pp.) 1s.
Standard I. (92 pp.)	6d.	Standard V. (216 pp.) 1s. 3d.
Standard II. (124 pp.)	9d.	Standard VI. (228 pp.) 1s. 6d.

* This Series has been abridged from "The Globe Readers" to meet the demand for smaller reading books.

GOLDSMITH.—THE TRAVELLER, or a Prospect of Society; and the DESERTED VILLAGE. With Notes, Philological and Explanatory, by J. W. HALES, M.A. Cr. 8vo. 6d.

THE VICAR OF WAKEFIELD. With a Memoir of Goldsmith, by Prof. MASSON. Gl. 8vo. 1s. (Globe Readings from Standard Authors.)

SELECT ESSAYS. With Introduction and Notes, by Prof. C. D. YONGE. Fcap. 8vo. 2s. 6d.

THE TRAVELLER AND THE DESERTED VILLAGE. With Introduction and Notes. By A. BARRETT, B.A., Professor of English Literature, Elphinstone College, Bombay. Gl. 8vo. 1s. 6d.

THE VICAR OF WAKEFIELD. With Introduction and Notes. By H. LITTLE-DALE, B.A., Professor of History and English Literature, Baroda College. Gl. 8vo. [In preparation.

GOSSE.—A HISTORY OF EIGHTEENTH CENTURY LITERATURE (1660-1780). By EDMUND GOSSE, M.A. Cr. 8vo. 7s. 6d.

GRAY.—POEMS. With Introduction and Notes, by JOHN BRADSHAW, LL.D. Gl. 8vo. [In preparation.

HALES.—LONGER ENGLISH POEMS. With Notes, Philological and Explanatory, and an Introduction on the Teaching of English, by J. W. HALES, M.A., Professor of English Literature at King's College, London. Ex. fcap. 8vo. 4s. 6d.

HELPS.—ESSAYS WRITTEN IN THE INTERVALS OF BUSINESS. With Introduction and Notes, by F. J. ROWE, M.A., and W. T. WEBB, M.A., Gl. 8vo. 2s. 6d.

JOHNSON.—LIVES OF THE POETS. The Six Chief Lives (Milton, Dryden, Swift, Addison, Pope, Gray), with Macaulay's "Life of Johnson." With Preface and Notes by MATTHEW ARNOLD. Cr. 8vo. 4s. 6d.

LAMB.—TALES FROM SHAKSPEARE. With Preface by the Rev. CANON AINGER, M.A., LL.D. Gl. 8vo. 2s. (*Globe Readings from Standard Authors.*)

LITERATURE PRIMERS.—Edited by JOHN RICHARD GREEN, LL.D. 18mo. 1s. each.

ENGLISH GRAMMAR. By Rev. R. MORRIS, LL.D.

ENGLISH GRAMMAR EXERCISES. By R. MORRIS, LL.D., and H. C. BOWEN, M.A.

EXERCISES ON MORRIS'S PRIMER OF ENGLISH GRAMMAR. By J WETHERELL, M.A.

ENGLISH COMPOSITION. By Professor NICHOL.

QUESTIONS AND EXERCISES ON ENGLISH COMPOSITION By Prof. NICHOL and W. S. M'CORMICK.

ENGLISH LITERATURE. By STOPFORD BROOKE, M.A.

SHAKSPERE. By Professor DOWDEN.

THE CHILDREN'S TREASURY OF LYRICAL POETRY. Selected and arranged with Notes by FRANCIS TURNER PALGRAVE. In Two Parts. 1s. each.

PHILOLOGY. By J. PEILE, Litt.D.

ROMAN LITERATURE. By Prof. A. S. WILKINS, Litt.D.

GREEK LITERATURE. By Prof. JEBB, Litt.D.

HOMER. By the Rt. Hon. W. E. GLADSTONE, M.P.

A HISTORY OF ENGLISH LITERATURE IN FOUR VOLUMES. Cr. 8vo.

EARLY ENGLISH LITERATURE. By STOPFORD BROOKE, M.A. [*In preparation.*

ELIZABETHAN LITERATURE. (1560-1665.) By GEORGE SAINTSBURY. 7s. 6d.

EIGHTEENTH CENTURY LITERATURE. (1660-1780.) By EDMUND GOSSE. M.A. 7s. 6d.

THE MODERN PERIOD. By Prof. DOWDEN. [*In preparation.*

MACMILLAN'S READING BOOKS.

PRIMER. 18mo. 48 pp. 2d.	BOOK IV. for Standard IV. 176 pp. 8d.
BOOK I. for Standard I. 96 pp. 4d.	
BOOK II. for Standard II. 144 pp. 5d.	BOOK V. for Standard V. 380 pp. 1s.
BOOK III. for Standard III. 160 pp. 6d.	BOOK VI. for Standard VI. Cr. 8vo. 430 pp. 2s.

Book VI. is fitted for Higher Classes, and as an Introduction to English Literature.

MACMILLAN'S COPY BOOKS.—1. Large Post 4to. Price 4d. each. 2. Post Oblong. Price 2d. each.

1. INITIATORY EXERCISES AND SHORT LETTERS.
2. WORDS CONSISTING OF SHORT LETTERS.
*3. LONG LETTERS. With Words containing Long Letters—Figures.
*4. WORDS CONTAINING LONG LETTERS.
4a. PRACTISING AND REVISING COPY-BOOK. For Nos. 1 to 4.
*5. CAPITALS AND SHORT HALF-TEXT. Words beginning with a Capital.
*6. HALF-TEXT WORDS beginning with Capitals—Figures.
*7. SMALL-HAND AND HALF-TEXT. With Capitals and Figures.
*8. SMALL-HAND AND HALF-TEXT. With Capitals and Figures.
8a. PRACTISING AND REVISING COPY-BOOK. For Nos. 5 to 8.
*9. SMALL-HAND SINGLE HEADLINES—Figures.
10. SMALL-HAND SINGLE HEADLINES—Figures.
11. SMALL-HAND DOUBLE HEADLINES—Figures.
12. COMMERCIAL AND ARITHMETICAL EXAMPLES, &c.
12a. PRACTISING AND REVISING COPY-BOOK. For Nos. 8 to 12.
* *These numbers may be had with Goodman's Patent Sliding Copies.* Large Post 4to. Price 6d. each.

MARTIN.—THE POET'S HOUR: Poetry selected and arranged for Children. By FRANCES MARTIN. 18mo. 2s. 6d.

SPRING-TIME WITH THE POETS. By the same. 18mo. 3s. 6d.

MILTON.—PARADISE LOST. Books I. and II. With Introduction and Notes, by MICHAEL MACMILLAN, B.A., Professor of Logic and Moral Philosophy, Elphinstone College, Bombay. Gl. 8vo. 2s. 6d. Or separately, 1s. 6d. each.

L'ALLEGRO, IL PENSEROSO, LYCIDAS, ARCADES, SONNETS, &c. With Introduction and Notes, by W. BELL, M.A., Professor of Philosophy and Logic, Government College, Lahore. Gl. 8vo. 2s.

COMUS. By the same. Gl. 8vo. 1s. 6d.

SAMSON AGONISTES. By H. M. PERCIVAL, M.A., Professor of English Literature, Presidency College, Calcutta. Gl. 8vo. 2s. 6d.

INTRODUCTION TO THE STUDY OF MILTON. By STOPFORD BROOKE, M.A. Fcap. 8vo. 1s. 6d. (*Classical Writers.*)

MORLEY.—ON THE STUDY OF LITERATURE. Address to the Students of the London Society for the Extension of University Teaching, delivered at the Mansion House, February 26, 1887. By JOHN MORLEY. Gl. 8vo, cloth. 1s. 6d.
* Also a Popular Edition in Pamphlet form for Distribution, price 2d.

APHORISMS. Address delivered before the Philosophical Society of Edinburgh, November 11, 1887. By the same. Gl. 8vo. 1s. 6d.

MORRIS.—Works by the Rev. R. MORRIS, LL.D.

PRIMER OF ENGLISH GRAMMAR. 18mo. 1s.

ELEMENTARY LESSONS IN HISTORICAL ENGLISH GRAMMAR, containing Accidence and Word Formation. 18mo. 2s. 6d.

HISTORICAL OUTLINES OF ENGLISH ACCIDENCE, comprising Chapters on the History and Development of the Language, and on Word Formation. Ex. fcap. 8vo. 6s.

MORRIS AND KELLNER.—HISTORICAL OUTLINES OF ENGLISH SYNTAX. By Rev. R. MORRIS and Dr. L. KELLNER.   [*In preparation.*

NICHOL.—A SHORT HISTORY OF ENGLISH LITERATURE. By Prof. JOHN NICHOL. Gl. 8vo.   [*In preparation.*

OLIPHANT.—THE OLD AND MIDDLE ENGLISH. By T. L. KINGTON OLIPHANT. New Ed., revised and enlarged, of "The Sources of Standard English." Gl. 8vo. 9s.

THE NEW ENGLISH. By the same. 2 vols. Cr. 8vo. 21s.

PALGRAVE.—THE CHILDREN'S TREASURY OF LYRICAL POETRY. Selected and arranged, with Notes, by FRANCIS T. PALGRAVE. 18mo. 2s. 6d. Also in Two Parts. 1s. each.

PATMORE.—THE CHILDREN'S GARLAND FROM THE BEST POETS. Selected and arranged by COVENTRY PATMORE. Gl. 8vo. 2s. (*Globe Readings from Standard Authors.*)

PLUTARCH.—Being a Selection from the Lives which illustrate Shakespeare. North's Translation. Edited, with Introductions, Notes, Index of Names, and Glossarial Index, by Prof. W. W. SKEAT, Litt.D. Cr. 8vo. 6s.

RANSOME.—SHORT STUDIES OF SHAKESPEARE'S PLOTS. By CYRIL RANSOME, Professor of Modern History and Literature, Yorkshire College, Leeds. Cr. 8vo. 3s. 6d.

RYLAND.—CHRONOLOGICAL OUTLINES OF ENGLISH LITERATURE. By F. RYLAND, M.A. Cr. 8vo.   [*In the Press.*

SAINTSBURY.—A HISTORY OF ELIZABETHAN LITERATURE. 1560-1665. By GEORGE SAINTSBURY. Cr. 8vo. 7s. 6d.

SCOTT.—LAY OF THE LAST MINSTREL, and THE LADY OF THE LAKE. Edited, with Introduction and Notes, by FRANCIS TURNER PALGRAVE. Gl. 8vo. 1s. (*Globe Readings from Standard Authors.*)

THE LAY OF THE LAST MINSTREL. With Introduction and Notes, by G. H. STUART, M.A., Professor of English Literature, Presidency College, Madras. Gl. 8vo. Cantos I. to III. 1s. 6d. Introduction and Canto I. 9d.

MARMION, and THE LORD OF THE ISLES. By F. T. PALGRAVE. Gl. 8vo. 1s. (*Globe Readings from Standard Authors.*)

MARMION. With Introduction and Notes, by MICHAEL MACMILLAN, B.A. Gl. 8vo. 3s. 6d.

THE LADY OF THE LAKE. By G. H. STUART, M.A.   [*In the Press.*

ROKEBY. With Introduction and Notes, by MICHAEL MACMILLAN, B.A. Gl. 8vo. 3s. 6d.

SHAKESPEARE.—A SHAKESPEARIAN GRAMMAR. By Rev. E. A. ABBOTT, D.D. Gl. 8vo. 6s.

A SHAKESPEARE MANUAL. By F. G. FLEAY, M.A. 2d Ed. Ex. fcap. 8vo. 4s. 6d.

PRIMER OF SHAKESPERE. By Prof. DOWDEN. 18mo. 1s.

SHORT STUDIES OF SHAKESPEARE'S PLOTS. By CYRIL RANSOME, M.A. Cr. 8vo. 3s. 6d.

THE TEMPEST. With Introduction and Notes, by K. DEIGHTON, late Principal of Agra College. Gl. 8vo. 1s. 6d.

MUCH ADO ABOUT NOTHING. By the same. Gl. 8vo. 2s.

THE MERCHANT OF VENICE. By the same. Gl. 8vo. 1s. 6d.

TWELFTH NIGHT. By the same. Gl. 8vo. 1s. 6d.

THE WINTER'S TALE. By the same. Gl. 8vo. 2s. 6d.

RICHARD II. By the same. Gl. 8vo. [In August

KING JOHN. By the same. Gl. 8vo. [In preparation.

HENRY V. By the same. Gl. 8vo. 2s.

RICHARD III. By C. H. TAWNEY, M.A., Principal and Professor of English Literature, Presidency College, Calcutta. Gl. 8vo. 2s. 6d.

JULIUS CÆSAR. By K. DEIGHTON. Gl. 8vo. 2s.

MACBETH. By the same. Gl. 8vo. 1s. 6d.

OTHELLO. By the same. Gl. 8vo. 2s. 6d.

CYMBELINE. By the same. Gl. 8vo. 2s. 6d.

SONNENSCHEIN AND MEIKLEJOHN.—THE ENGLISH METHOD OF TEACHING TO READ. By A. SONNENSCHEIN and J. M. D. MEIKLEJOHN, M.A. Fcap. 8vo.

COMPRISING:

THE NURSERY BOOK, containing all the Two-Letter Words in the Language. 1d. (Also in Large Type on Sheets for School Walls. 5s.)

THE FIRST COURSE, consisting of Short Vowels with Single Consonants. 7d.

THE SECOND COURSE, with Combinations and Bridges, consisting of Short Vowels with Double Consonants. 7d.

THE THIRD AND FOURTH COURSES, consisting of Long Vowels, and all the Double Vowels in the Language. 7d.

SOUTHEY.—LIFE OF NELSON. With Introduction and Notes, by MICHAEL MACMILLAN, B.A. Gl. 8vo. 3s. 6d.

TAYLOR.—WORDS AND PLACES; or, Etymological Illustrations of History, Ethnology, and Geography. By Rev. ISAAC TAYLOR, Litt.D. With Maps. Gl. 8vo. 6s.

TENNYSON.—THE COLLECTED WORKS OF LORD TENNYSON. An Edition for Schools. In Four Parts. Cr. 8vo. 2s. 6d. each.

TENNYSON FOR THE YOUNG. Edited, with Notes for the Use of Schools, by the Rev. ALFRED AINGER, LL.D., Canon of Bristol. [In preparation.

SELECTIONS FROM TENNYSON. With Introduction and Notes, by F. J. ROWE, M.A., and W. T. WEBB, M.A. Gl. 8vo. 3s. 6d.

This selection contains :—Recollections of the Arabian Nights, The Lady of Shalott, Oenone, The Lotos Eaters, Ulysses, Tithonus, Morte d'Arthur, Sir Galahad, Dora, Ode on the Death of the Duke of Wellington, and The Revenge.

THRING.—THE ELEMENTS OF GRAMMAR TAUGHT IN ENGLISH. By EDWARD THRING, M.A. With Questions. 4th Ed. 18mo. 2s.

VAUGHAN.—WORDS FROM THE POETS. By C. M. VAUGHAN. 18mo. 1s.

WARD.—THE ENGLISH POETS. Selections, with Critical Introductions by various Writers and a General Introduction by MATTHEW ARNOLD. Edited by T. H. WARD, M.A. 4 Vols. Vol. I. CHAUCER TO DONNE.—Vol. II. BEN JONSON TO DRYDEN.—Vol. III. ADDISON TO BLAKE.—Vol. IV. WORDSWORTH TO ROSSETTI. Cr. 8vo. Each 7s. 6d.

B

WETHERELL.—EXERCISES ON MORRIS'S PRIMER OF ENGLISH GRAM-MAR. By JOHN WETHERELL, M.A., Headmaster of Towcester Grammar School. 18mo. 1s.

WOODS.—A FIRST POETRY BOOK. By M. A. WOODS, Head Mistress of the Clifton High School for Girls. Fcap. 8vo. 2s. 6d.

A SECOND POETRY BOOK By the same. In two Parts. 2s. 6d. each.

A THIRD POETRY BOOK. By the same. 4s. 6d.

WORDSWORTH.—SELECTIONS. With Introduction and Notes, by WILLIAM WORDSWORTH, B.A., Principal and Professor of History and Political Economy, Elphinstone College, Bombay   Gl. 8vo.   [*In preparation.*

YONGE.—A BOOK OF GOLDEN DEEDS. By CHARLOTTE M. YONGE. Gl. 8vo. 2s.

THE ABRIDGED BOOK OF GOLDEN DEEDS. 18mo. 1s.

FRENCH.

BEAUMARCHAIS.—LE BARBIER DE SEVILLE. With Introduction and Notes. By L. P. BLOUET. Fcap. 8vo. 3s. 6d.

BOWEN.—FIRST LESSONS IN FRENCH. By H. COURTHOPE BOWEN, M.A. Ex. fcap. 8vo. 1s.

BREYMANN.—Works by HERMANN BREYMANN, Ph.D., Professor of Philology in the University of Munich.

A FRENCH GRAMMAR BASED ON PHILOLOGICAL PRINCIPLES. Ex. fcap. 8vo. 4s. 6d.

FIRST FRENCH EXERCISE BOOK. Ex. fcap. 8vo. 4s. 6d.

SECOND FRENCH EXERCISE BOOK. Ex. fcap. 8vo. 2s. 6d.

FASNACHT.—Works by G. E. FASNACHT, late Assistant Master at Westminster.

THE ORGANIC METHOD OF STUDYING LANGUAGES. Ex. fcap. 8vo. I. French. 3s. 6d.

A SYNTHETIC FRENCH GRAMMAR FOR SCHOOLS. Cr. 8vo. 3s. 6d.

GRAMMAR AND GLOSSARY OF THE FRENCH LANGUAGE OF THE SEVENTEENTH CENTURY. Cr. 8vo.   [*In preparation.*

MACMILLAN'S PRIMARY SERIES OF FRENCH READING BOOKS.—Edited by G. E. FASNACHT. With Illustrations, Notes, Vocabularies, and Exercises. Gl. 8vo.

CORNAZ—NOS ENFANTS ET LEURS AMIS. By EDITH HARVEY. 1s. 6d.

DE MAISTRE—LA JEUNE SIBÉRIENNE ET LE LÉPREUX DE LA CITÉ D'AOSTE. By STEPHANE BARLET, B.Sc. &c. 1s. 6d.

FLORIAN—FABLES. By Rev. CHARLES YELD, M.A., Headmaster of University School, Nottingham. 1s. 6d.

LA FONTAINE—A SELECTION OF FABLES. By L. M. MORIARTY, B.A., Assistant Master at Harrow. 2s. 6d.

MOLESWORTH—FRENCH LIFE IN LETTERS. By Mrs. MOLESWORTH. 1s. 6d.

PERRAULT—CONTES DE FÉES. By G. E. FASNACHT. 1s. 6d.

MACMILLAN'S PROGRESSIVE FRENCH COURSE.—By G. E. FASNACHT. Ex. fcp. 8vo.

FIRST YEAR, containing Easy Lessons on the Regular Accidence. 1s.

SECOND YEAR, containing an Elementary Grammar with copious Exercises, Notes, and Vocabularies. 2s.

THIRD YEAR, containing a Systematic Syntax, and Lessons in Composition. 2s. 6d.

THE TEACHER'S COMPANION TO MACMILLAN'S PROGRESSIVE FRENCH COURSE. With Copious Notes, Hints for Different Renderings, Synonyms, Philological Remarks, etc. By G. E. FASNACHT. Ex. fcap. 8vo. Each Year 4s. 6d.

MACMILLAN'S FRENCH COMPOSITION.—By G. E. FASNACHT. Ex. fcap. 8vo. Part I. Elementary. 2s. 6d. Part II. Advanced.   [*In the Press.*

THE TEACHER'S COMPANION TO MACMILLAN'S COURSE OF FRENCH COMPOSITION. By G. E. FASNACHT. Part I. Ex. fcap. 8vo. 4s. 6d.

MACMILLAN'S PROGRESSIVE FRENCH READERS. By G. E. FASNACHT. Ex. fcap. 8vo.

FIRST YEAR, containing Tales, Historical Extracts, Letters, Dialogues, Ballads, Nursery Songs, etc., with Two Vocabularies: (1) in the order of subjects; (2) in alphabetical order. With Imitative Exercises. 2s. 6d.

SECOND YEAR, containing Fiction in Prose and Verse, Historical and Descriptive Extracts, Essays, Letters, Dialogues, etc. With Imitative Exercises. 2s. 6d.

MACMILLAN'S FOREIGN SCHOOL CLASSICS. Edited by G. E. FASNACHT. 18mo.

CORNEILLE—LE CID. By G. E. FASNACHT. 1s.

DUMAS—LES DEMOISELLES DE ST. CYR. By VICTOR OGER, Lecturer at University College, Liverpool. 1s. 6d.

LA FONTAINE'S FABLES. Books I.—VI. By L. M. MORIARTY, B.A., Assistant Master at Harrow. [In preparation.

MOLIÈRE—L'AVARE. By the same. 1s.

MOLIÈRE—LE BOURGEOIS GENTILHOMME. By the same. 1s. 6d.

MOLIÈRE—LES FEMMES SAVANTES. By G. E. FASNACHT. 1s.

MOLIÈRE—LE MISANTHROPE. By the same. 1s.

MOLIÈRE—LE MÉDECIN MALGRE LUI. By the same. 1s.

RACINE—BRITANICUS. By E. PELLISSIER, M.A., Assistant Master at Clifton College. 2s.

FRENCH READINGS FROM ROMAN HISTORY. Selected from various Authors, by C. COLBECK, M.A., Assistant Master at Harrow. 4s. 6d.

SAND, GEORGE—LA MARE AU DIABLE. By W. E. RUSSELL, M.A., Assistant Master at Haileybury. 1s.

SANDEAU, JULES—MADEMOISELLE DE LA SEIGLIERE. By H. C. STEEL, Assistant Master at Winchester. 1s. 6d.

THIERS'S HISTORY OF THE EGYPTIAN EXPEDITION. By Rev. H. A. BULL, M.A., Assistant Master at Wellington. [In preparation.

VOLTAIRE—CHARLES XII. By G. E. FASNACHT. 3s. 6d.

MASSON.—A COMPENDIOUS DICTIONARY OF THE FRENCH LANGUAGE. Adapted from the Dictionaries of Professor A. ELWALL. By GUSTAVE MASSON. Cr. 8vo. 6s.

MOLIERE.—LE MALADE IMAGINAIRE. With Introduction and Notes, by F. TARVER, M.A., Assistant Master at Eton. Fcap. 8vo. 2s. 6d.

PELLISSIER.—FRENCH ROOTS AND THEIR FAMILIES. A Synthetic Vocabulary, based upon Derivations. By E. PELLISSIER, M.A., Assistant Master at Clifton College. Gl. 8vo. 6s.

GERMAN.

HUSS.—A SYSTEM OF ORAL INSTRUCTION IN GERMAN, by means of Progressive Illustrations and Applications of the leading Rules of Grammar. By H. C. O. HUSS, Ph.D. Cr. 8vo. 5s.

MACMILLAN'S PROGRESSIVE GERMAN COURSE. By G. E. FASNACHT. Ex. fcp. 8vo.

FIRST YEAR. Easy lessons and Rules on the Regular Accidence. 1s. 6d.

SECOND YEAR. Conversational Lessons in Systematic Accidence and Elementary Syntax. With Philological Illustrations and Etymological Vocabulary. 3s. 6d.

THIRD YEAR. [In the Press.

TEACHER'S COMPANION TO MACMILLAN'S PROGRESSIVE GERMAN COURSE. With copious Notes, Hints for Different Renderings, Synonyms, Philological Remarks, etc. By G. E. FASNACHT. Ex. fcap. 8vo. FIRST YEAR. 4s. 6d. SECOND YEAR. 4s. 6d.

MACMILLAN'S PROGRESSIVE GERMAN READERS. By G. E. FASNACHT. Ex. fcap. 8vo.

FIRST YEAR, containing an Introduction to the German order of Words, with Copious Examples, extracts from German Authors in Prose and Poetry; Notes, and Vocabularies. 2s. 6d.

MACMILLAN'S PRIMARY SERIES OF GERMAN READING BOOKS. Edited by G. E. FASNACHT. With Notes, Vocabularies and Exercises. Gl. 8vo.

GRIMM—KINDER UND HAUSMÄRCHEN. By G. E. FASNACHT. 2s. 6d.

HAUFF—DIE KARAVANE. By HERMAN HAGER, Ph.D., Lecturer in the Owens College, Manchester. 3s.

SCHMID, CHR. VON—H. VON EICHENFELS. By G. E. FASNACHT. 2s. 6d.

MACMILLAN'S FOREIGN SCHOOL CLASSICS.—Edited by G. E. FASNACHT. 18mo.

FREYTAG (G.).—DOKTOR LUTHER. By F. STORR, M.A., Headmaster of the Modern Side, Merchant Taylors' School.   [In preparation.

GOETHE—GÖTZ VON BERLICHINGEN. By H. A. BULL, M.A., Assistant Master at Wellington. 2s.

GOETHE—FAUST. PART I., followed by an Appendix on PART II. By JANE LEE, Lecturer in German Literature at Newnham College, Cambridge. 4s. 6d.

HEINE—SELECTIONS FROM THE REISEBILDER AND OTHER PROSE WORKS. By C. COLBECK, M.A., Assistant Master at Harrow. 2s. 6d.

LESSING—MINNA VON BARNHELM. By JAMES SIME, M.A. [In preparation.

SCHILLER—SELECTIONS FROM SCHILLER'S LYRICAL POEMS. With a Memoir of Schiller. By E. J. TURNER, B.A., and E. D. A. MORSHEAD, M.A. Assistant Masters at Winchester. 2s. 6d.

SCHILLER—DIE JUNGFRAU VON ORLEANS. By JOSEPH GOSTWICK. 2s. 6d.

SCHILLER—MARIA STUART. By C. SHELDON, D.Lit., of the Royal Academical Institution, Belfast. 2s. 6d.

SCHILLER—WILHELM TELL. By G. E. FASNACHT. 2s. 6d.

SCHILLER—WALLENSTEIN. Part I. DAS LAGER. By H. B. COTTERILL, M.A. 2s.

UHLAND—SELECT BALLADS. Adapted as a First Easy Reading Book for Beginners. With Vocabulary. By G. E. FASNACHT. 1s.

PYLODET.—NEW GUIDE TO GERMAN CONVERSATION ; containing an Alphabetical List of nearly 800 Familiar Words ; followed by Exercises, Vocabulary of Words in frequent use, Familiar Phrases and Dialogues, a Sketch of German Literature, Idiomatic Expressions, etc. By L. PYLODET. 18mo. 2s. 6d.

WHITNEY.—A COMPENDIOUS GERMAN GRAMMAR. By W. D. WHITNEY, Professor of Sanskrit and Instructor in Modern Languages in Yale College. Cr. 8vo. 4s. 6d.

A GERMAN READER IN PROSE AND VERSE. By the Same. With Notes and Vocabulary. Cr. 8vo. 5s.

WHITNEY AND EDGREN.—A COMPENDIOUS GERMAN AND ENGLISH DICTIONARY, with Notation of Correspondences and Brief Etymologies. By Prof. W. D. WHITNEY, assisted by A. H. EDGREN. Cr. 8vo. 7s. 6d.

THE GERMAN-ENGLISH PART, separately, 5s.

MODERN GREEK.

VINCENT AND DICKSON.—HANDBOOK TO MODERN GREEK. By Sir EDGAR VINCENT, K.C.M.G., and T. G. DICKSON, M.A. With Appendix on the relation of Modern and Classical Greek by Prof. JEBB. Cr. 8vo. 6s.

ITALIAN.

DANTE.—THE PURGATORY OF DANTE. With Translation and Notes, by A. J. BUTLER, M.A. Cr. 8vo. 12s. 6d.

THE PARADISO OF DANTE. With Translation and Notes, by the Same. Cr. 8vo. 12s. 6d.

READINGS ON THE PURGATORIO OF DANTE. Chiefly based on the Commentary of Benvenuto Da Imola. By the Hon. W. WARREN VERNON, M.A. With an Introduction by the Very Rev the DEAN OF ST. PAUL'S. 2 vols. Cr. 8vo. 24s.

SPANISH.

CALDERON.—FOUR PLAYS OF CALDERON. With Introduction and Notes.
By NORMAN MACCOLL, M.A. Cr. 8vo. 14s.
  The four plays here given are *El Principe Constante, La Vida es Sueno, El Alcalde de Zalamea,* and *El Escondido y La Tapada.*

MATHEMATICS.

Arithmetic, Book-keeping, Algebra, Euclid and Pure Geometry, Geometrical
Drawing, Mensuration, Trigonometry, Analytical Geometry (Plane and
Solid), Problems and Questions in Mathematics, Higher Pure Mathematics, Mechanics (Statics, Dynamics, Hydrostatics, Hydrodynamics: see
also Physics), Physics (Sound, Light, Heat, Electricity, Elasticity, Attrac-
tions, &c.), Astronomy, Historical.

ARITHMETIC.

ALDIS.—THE GREAT GIANT ARITHMOS. A most Elementary Arithmetic
for Children. By MARY STEADMAN ALDIS. Illustrated. Gl. 8vo. 2s. 6d.
ARMY PRELIMINARY EXAMINATION, SPECIMENS OF PAPERS SET AT
THE, 1882-89.—With Answers to the Mathematical Questions. Subjects:
Arithmetic, Algebra, Euclid, Geometrical Drawing, Geography, French,
English Dictation. Cr. 8vo. 3s. 6d.
BRADSHAW.—A COURSE OF EASY ARITHMETICAL EXAMPLES FOR
BEGINNERS. By J. G. BRADSHAW, B.A., Assistant Master at Clifton College.
Gl. 8vo. 2s. With Answers, 2s. 6d.
BROOKSMITH.—ARITHMETIC IN THEORY AND PRACTICE. By J. BROOK-
SMITH, M.A. Cr. 8vo. 4s. 6d.
BROOKSMITH.—ARITHMETIC FOR BEGINNERS. By J. and E. J. BROOK-
SMITH. Gl. 8vo. 1s. 6d.
CANDLER.—HELP TO ARITHMETIC. Designed for the use of Schools. By H.
CANDLER, Mathematical Master of Uppingham School. 2d Ed. Ex. fcap. 8vo.
2s. 6d.
DALTON.—RULES AND EXAMPLES IN ARITHMETIC. By the Rev. T. DAL-
TON, M.A., Assistant Master at Eton. New Ed., with Answers. 18mo. 2s. 6d.
GOYEN—HIGHER ARITHMETIC AND ELEMENTARY MENSURATION.
By P. GOYEN, Inspector of Schools, Dunedin, New Zealand. Cr. 8vo. 5s.
HALL AND KNIGHT.—ARITHMETICAL EXERCISES AND EXAMINATION
PAPERS. With an Appendix containing Questions in LOGARITHMS and
MENSURATION. By H. S. HALL, M.A., Master of the Military and Engineering
Side, Clifton College, and S. R. KNIGHT, B.A. Gl. 8vo. 2s. 6d.
LOCK.—Works by Rev. J. B. LOCK, M.A., Senior Fellow, Assistant Tutor and
Lecturer in Gonville and Caius College, Cambridge.
  ARITHMETIC FOR SCHOOLS. With Answers and 1000 additional Examples
for Exercise. 3d Ed., revised. Gl. 8vo. 4s. 6d. Or in Two Parts:—
Part I. Up to and including Practice. 2s. Part II. With 1000 additional
Examples for Exercise. 3s. KEY. Cr. 8vo. 10s. 6d.
  ARITHMETIC FOR BEGINNERS. A School Class-Book of Commercial Arith-
metic. Gl. 8vo. 2s. 6d. KEY. Cr. 8vo. 8s. 6d.
  A SHILLING CLASS-BOOK OF ARITHMETIC, ADAPTED FOR USE IN
ELEMENTARY SCHOOLS. 18mo. 1s. With Answers 1s. 6d. [*In July.*

PEDLEY.—EXERCISES IN ARITHMETIC for the Use of Schools. Containing more than 7000 original Examples. By SAMUEL PEDLEY. Cr. 8vo. 5s. Also in Two Parts, 2s. 6d. each.

SMITH.—Works by Rev. BARNARD SMITH, M.A., late Fellow and Senior Bursar of St. Peter's College, Cambridge.

ARITHMETIC AND ALGEBRA, in their Principles and Application; with numerous systematically arranged Examples taken from the Cambridge Examination Papers, with especial reference to the Ordinary Examination for the B.A. Degree. New Ed., carefully revised. Cr. 8vo. 10s. 6d.

ARITHMETIC FOR SCHOOLS. Cr. 8vo. 4s. 6d. KEY. Cr. 8vo. 4s. 6d.

EXERCISES IN ARITHMETIC. Cr. 8vo. 2s. With Answers, 2s. 6d. Answers separately, 6d.

SCHOOL CLASS-BOOK OF ARITHMETIC. 18mo. Or separately, in Three Parts, 1s. each. KEYS. Parts I., II., and III., 2s. 6d. each.

SHILLING BOOK OF ARITHMETIC. 18mo. Or separately, Part I., 2d.; Part II., 3d.; Part III., 7d. Answers, 6d. KEY. 18mo. 4s. 6d.

THE SAME, with Answers. 18mo, cloth. 1s. 6d.

EXAMINATION PAPERS IN ARITHMETIC. 18mo. 1s. 6d. The Same, with Answers. 18mo. 2s. Answers, 6d. KEY. 18mo. 4s. 6d.

THE METRIC SYSTEM OF ARITHMETIC, ITS PRINCIPLES AND APPLICATIONS, with Numerous Examples. 18mo. 3d.

A CHART OF THE METRIC SYSTEM, on a Sheet, size 42 in. by 34 in. on Roller. 3s. 6d. Also a Small Chart on a Card. Price 1d.

EASY LESSONS IN ARITHMETIC, combining Exercises in Reading, Writing, Spelling, and Dictation. Part I. Cr. 8vo. 9d.

EXAMINATION CARDS IN ARITHMETIC. With Answers and Hints.

Standards I. and II., in box, 1s. Standards III., IV., and V., in boxes, 1s. each. Standard VI. in Two Parts, in boxes, 1s. each.

A and B papers, of nearly the same difficulty, are given so as to prevent copying, and the colours of the A and B papers differ in each Standard, and from those of every other Standard, so that a master or mistress can see at a glance whether the children have the proper papers.

BOOK-KEEPING.

THORNTON.—FIRST LESSONS IN BOOK-KEEPING. By J. THORNTON. Cr. 8vo. 2s. 6d. KEY. Oblong 4to. 10s. 6d.

PRIMER OF BOOK-KEEPING. 18mo. 1s. KEY.   [Immediately.

ALGEBRA.

DALTON.—RULES AND EXAMPLES IN ALGEBRA. By Rev. T. DALTON, Assistant Master at Eton. Part I. 18mo. 2s. KEY. Cr. 8vo. 7s. 6d. Part II. 18mo. 2s. 6d.

HALL AND KNIGHT.—Works by H. S. HALL, M.A., Master of the Military and Engineering Side, Clifton College, and S. R. KNIGHT, B.A.

ELEMENTARY ALGEBRA FOR SCHOOLS. 5th Ed., revised and corrected. Gl. 8vo, bound in maroon coloured cloth, 3s. 6d.; with Answers, bound in green coloured cloth, 4s. 6d.   [KEY. In the Press.

ALGEBRAICAL EXERCISES AND EXAMINATION PAPERS. To accompany ELEMENTARY ALGEBRA. 2d Ed., revised. Gl. 8vo. 2s. 6d.

HIGHER ALGEBRA. 3d Ed. Cr. 8vo. 7s. 6d. KEY. Cr. 8vo. 10s. 6d.

JONES AND CHEYNE.—ALGEBRAICAL EXERCISES. Progressively Arranged. By Rev. C. A. JONES and C. H. CHEYNE, M.A., late Mathematical Masters at Westminster School. 18mo. 2s. 6d.

KEY. By Rev. W. FAILES, M.A., Mathematical Master at Westminster School. Cr. 8vo. 7s. 6d.

SMITH.—ARITHMETIC AND ALGEBRA, in their Principles and Application; with numerous systematically arranged Examples taken from the Cambridge Examination Papers, with especial reference to the Ordinary Examination for the B.A. Degree. By Rev. BARNARD SMITH, M.A. New Edition, carefully revised. Cr. 8vo. 10s. 6d.

SMITH.—Works by CHARLES SMITH, M.A., Master of Sidney Sussex College, Cambridge.
  ELEMENTARY ALGEBRA. 2d Ed., revised. Gl. 8vo. 4s. 6d.
  A TREATISE ON ALGEBRA. Cr. 8vo. 7s. 6d. KEY. Cr. 8vo. 10s. 6d.

TODHUNTER.—Works by ISAAC TODHUNTER, F.R.S.
  ALGEBRA FOR BEGINNERS. 18mo. 2s. 6d. KEY. Cr. 8vo. 6s. 6d.
  ALGEBRA FOR COLLEGES AND SCHOOLS. Cr. 8vo. 7s. 6d. KEY. Cr. 8vo. 10s. 6d.

EUCLID AND PURE GEOMETRY.

COOKSHOTT AND WALTERS.—A TREATISE ON GEOMETRICAL CONICS. In accordance with the Syllabus of the Association for the Improvement of Geometrical Teaching. By A. COOKSHOTT, M.A., Assistant Master at Eton; and Rev. F. B. WALTERS, M.A., Principal of King William's College, Isle of Man. Cr. 8vo. 5s.

CONSTABLE.—GEOMETRICAL EXERCISES FOR BEGINNERS. By SAMUEL CONSTABLE. Cr. 8vo. 3s. 6d.

CUTHBERTSON.—EUCLIDIAN GEOMETRY. By FRANCIS CUTHBERTSON, M.A., LL.D. Ex. fcap. 8vo. 4s. 6d.

DAY.—PROPERTIES OF CONIC SECTIONS PROVED GEOMETRICALLY. By Rev. H. G. DAY, M.A. Part I. The Ellipse, with an ample collection of Problems. Cr. 8vo. 3s. 6d.

DODGSON.—Works by CHARLES L. DODGSON, M.A., Student and late Mathematical Lecturer, Christ Church, Oxford.
  EUCLID, BOOKS I. AND II. 6th Ed., with words substituted for the Algebraical Symbols used in the 1st Ed. Cr. 8vo. 2s.
  EUCLID AND HIS MODERN RIVALS. 2d Ed. Cr. 8vo. 6s.
  CURIOSA MATHEMATICA. Part I. A New Theory of Parallels. 2d Ed. Cr. 8vo. 2s.

DREW.—GEOMETRICAL TREATISE ON CONIC SECTIONS. By W. H. DREW, M.A. New Ed., enlarged. Cr. 8vo. 5s.

DUPUIS.—ELEMENTARY SYNTHETIC GEOMETRY OF THE POINT, LINE, AND CIRCLE IN THE PLANE. By N. F. DUPUIS, M.A., Professor of Pure Mathematics in the University of Queen's College, Kingston, Canada. Gl. 8vo. 4s. 6d.

HALL AND STEVENS.—A TEXT-BOOK OF EUCLID'S ELEMENTS. Including Alternative Proofs, together with additional Theorems and Exercises, classified and arranged. By H. S. HALL, M.A., and F. H. STEVENS, M.A., Masters of the Military and Engineering Side, Clifton College. Gl. 8vo. Book I., 1s.; Books I. and II., 1s. 6d.; Books I.–IV., 3s.; Books III.–VI., 3s.; Books I.–VI. and XI., 4s. 6d.; Book XI., 1s.   [KEY. In preparation.

HALSTED.—THE ELEMENTS OF GEOMETRY By G. B. HALSTED, Professor of Pure and Applied Mathematics in the University of Texas. 8vo. 12s. 6d.

LOCK.—EUCLID FOR BEGINNERS. Being an Introduction to existing Text-books. By Rev. J. B. LOCK, M.A.   [In the Press.

MAULT.—NATURAL GEOMETRY: an Introduction to the Logical Study of Mathematics. For Schools and Technical Classes. With Explanatory Models based upon the Tachymetrical works of Ed. Lagout. By A. MAULT. 18mo. 1s. Models to Illustrate the above, in Box, 12s. 6d.

MILNE AND DAVIS.—GEOMETRICAL CONICS. Part I. The Parabola. By Rev. J. J. MILNE, M.A., and R. F. DAVIS, M.A. Cr. 8vo.   [In the Press.

SYLLABUS OF PLANE GEOMETRY (corresponding to Euclid, Books I.–VI.)— Prepared by the Association for the Improvement of Geometrical Teaching. Cr. 8vo. 1s.

SYLLABUS OF MODERN PLANE GEOMETRY.—Prepared by the Association for the Improvement of Geometrical Teaching. Cr. 8vo. Sewed. 1s.

TODHUNTER.—THE ELEMENTS OF EUCLID. By I. TODHUNTER, F.R.S. 18mo. 3s. 6d. KEY. Cr. 8vo. 6s. 6d.

WILSON.—Works by Rev. J. M. WILSON, M.A., Head Master of Clifton College.

     ELEMENTARY GEOMETRY. BOOKS I.-V. Containing the Subjects of Euclid's first Six Books. Following the Syllabus of the Geometrical Association. Ex. fcap. 8vo. 4s. 6d.

     SOLID GEOMETRY AND CONIC SECTIONS. With Appendices on Transversals and Harmonic Division. Ex. fcap. 8vo. 3s. 6d.

GEOMETRICAL DRAWING.

EAGLES.—CONSTRUCTIVE GEOMETRY OF PLANE CURVES. By T. H. EAGLES, M.A., Instructor in Geometrical Drawing and Lecturer in Architecture at the Royal Indian Engineering College, Cooper's Hill. Cr. 8vo. 12s.

EDGAR AND PRITCHARD.—NOTE-BOOK ON PRACTICAL SOLID OR DESCRIPTIVE GEOMETRY. Containing Problems with help for Solutions. By J. H. EDGAR and G. S. PRITCHARD. 4th Ed., revised by A. MEEZE. Gl. 8vo. 4s. 6d.

KITCHENER.—A GEOMETRICAL NOTE-BOOK. Containing Easy Problems in Geometrical Drawing preparatory to the Study of Geometry. For the Use of Schools. By F. E. KITCHENER, M.A., Head Master of the Newcastle-under-Lyme High School. 4to. 2s.

MILLAR.—ELEMENTS OF DESCRIPTIVE GEOMETRY. By J. B. MILLAR, Civil Engineer, Lecturer on Engineering in the Victoria University, Manchester. 2d Ed. Cr. 8vo. 6s.

PLANT.—GEOMETRICAL DRAWING. By E. C. PLANT. Globe 8vo.
[*In preparation.*

MENSURATION.

STEVENS.—ELEMENTARY MENSURATION. With Exercises on the Mensuration of Plane and Solid Figures. By F. H. STEVENS, M.A. Gl. 8vo.
[*In preparation.*

TEBAY.—ELEMENTARY MENSURATION FOR SCHOOLS. By S. TEBAY. Ex. fcap. 8vo. 3s. 6d.

TODHUNTER.—MENSURATION FOR BEGINNERS. By ISAAC TODHUNTER, F.R.S. 18mo. 2s. 6d. KEY. By Rev. FR. L. McCARTHY. Cr. 8vo. 7s. 6d.

TRIGONOMETRY.

BEASLEY.—AN ELEMENTARY TREATISE ON PLANE TRIGONOMETRY. With Examples. By R. D. BEASLEY, M.A. 9th Ed., revised and enlarged. Cr. 8vo. 3s. 6d.

BOTTOMLEY.—FOUR-FIGURE MATHEMATICAL TABLES. Comprising Logarithmic and Trigonometrical Tables, and Tables of Squares, Square Roots, and Reciprocals. By J. T. BOTTOMLEY, M.A., Lecturer in Natural Philosophy in the University of Glasgow. 8vo. 2s. 6d.

HAYWARD.—THE ALGEBRA OF CO-PLANAR VECTORS AND TRIGONOMETRY. By R. B. HAYWARD, M.A., F.R.S., Assistant Master at Harrow.
[*In preparation.*

JOHNSON.—A TREATISE ON TRIGONOMETRY. By W. E. JOHNSON, M.A. late Scholar and Assistant Mathematical Lecturer at King's College, Cambridge. Cr. 8vo. 8s. 6d.

LOCK.—Works by Rev. J. B. LOCK, M.A., Senior Fellow, Assistant Tutor and Lecturer in Gonville and Caius College, Cambridge.

     TRIGONOMETRY FOR BEGINNERS, as far as the Solution of Triangles. 3d Ed. Gl. 8vo. 2s. 6d. KEY. Cr. 8vo. 6s. 6d.

     ELEMENTARY TRIGONOMETRY. 6th Ed. (in this edition the chapter on logarithms has been carefully revised). Gl. 8vo. 4s. 6d. KEY. Cr. 8vo. 8s. 6d.

HIGHER TRIGONOMETRY. 5th Ed. Gl. 8vo. 4s. 6d. Both Parts complete in One Volume. Gl. 8vo. 7s. 6d.

M'CLELLAND AND PRESTON.—A TREATISE ON SPHERICAL TRIGONO-METRY. With applications to Spherical Geometry and numerous Examples. By W. J. M'CLELLAND, M.A., Principal of the Incorporated Society's School, Santry, Dublin, and T. PRESTON, M.A. Cr. 8vo. 8s. 6d., or: Part I. To the End of Solution of Triangles, 4s. 6d. Part II., 5s.

PALMER.—TEXT-BOOK OF PRACTICAL LOGARITHMS AND TRIGONO-METRY. By J. H. PALMER, Headmaster, R.N., H.M.S. Cambridge, Devon-port. Gl. 8vo. 4s. 6d.

SNOWBALL.—THE ELEMENTS OF PLANE AND SPHERICAL TRIGONO-METRY. By J. C. SNOWBALL. 14th Ed. Cr. 8vo. 7s. 6d.

TODHUNTER.—Works by ISAAC TODHUNTER, F.R.S.
TRIGONOMETRY FOR BEGINNERS. 18mo. 2s. 6d. KEY. Cr. 8vo. 8s. 6d.
PLANE TRIGONOMETRY. Cr. 8vo. 5s. KEY. Cr. 8vo. 10s. 6d.
A TREATISE ON SPHERICAL TRIGONOMETRY. Cr. 8vo. 4s. 6d.

WOLSTENHOLME.—EXAMPLES FOR PRACTICE IN THE USE OF SEVEN-FIGURE LOGARITHMS. By JOSEPH WOLSTENHOLME, D.Sc., late Professor of Mathematics in the Royal Indian Engineering Coll., Cooper's Hill. 8vo. 5s.

ANALYTICAL GEOMETRY (Plane and Solid).

DYER.—EXERCISES IN ANALYTICAL GEOMETRY. By J. M. DYER, M.A., Assistant Master at Eton. Illustrated. Cr. 8vo. 4s. 6d.

FERRERS.—AN ELEMENTARY TREATISE ON TRILINEAR CO-ORDIN-ATES, the Method of Reciprocal Polars, and the Theory of Projectors. By the Rev. N. M. FERRERS, D.D., F.R.S., Master of Gonville and Caius College, Cambridge. 4th Ed., revised. Cr. 8vo. 6s. 6d.

FROST.—Works by PERCIVAL FROST, D.Sc., F.R.S., Fellow and Mathematical Lecturer at King's College, Cambridge.
AN ELEMENTARY TREATISE ON CURVE TRACING. 8vo. 12s.
SOLID GEOMETRY. 3d Ed. Demy 8vo. 16s.
HINTS FOR THE SOLUTION OF PROBLEMS in the Third Edition of SOLID GEOMETRY. 8vo. 8s. 6d.

HAYWARD.—THE ELEMENTS OF SOLID GEOMETRY. By R. B. HAYWARD, M.A., F.R.S. Gl. 8vo. [In the Press.

JOHNSON.—CURVE TRACING IN CARTESIAN CO-ORDINATES. By W. WOOLSEY JOHNSON, Professor of Mathematics at the U.S. Naval Academy, Annapolis, Maryland. Cr. 8vo. 4s. 6d.

PUCKLE.—AN ELEMENTARY TREATISE ON CONIC SECTIONS AND AL-GEBRAIC GEOMETRY. With Numerous Examples and Hints for their Sol-ution. By G. H. PUCKLE, M.A. 5th Ed., revised and enlarged. Cr. 8vo. 7s. 6d.

SMITH.—Works by CHARLES SMITH, M.A., Master of Sidney Sussex College, Cambridge.
CONIC SECTIONS. 7th Ed. Cr. 8vo. 7s. 6d.
SOLUTIONS TO CONIC SECTIONS. Cr. 8vo. 10s. 6d.
AN ELEMENTARY TREATISE ON SOLID GEOMETRY. 2d Ed. Cr. 8vo. 9s. 6d.

TODHUNTER.—Works by ISAAC TODHUNTER, F.R.S.
PLANE CO-ORDINATE GEOMETRY, as applied to the Straight Line and the Conic Sections. Cr. 8vo. 7s. 6d.
KEY. By C. W. BOURNE, M.A., Headmaster of King's College School. Cr. 8vo. 10s. 6d.
EXAMPLES OF ANALYTICAL GEOMETRY OF THREE DIMENSIONS. New Ed., revised. Cr. 8vo. 4s.

C

PROBLEMS AND QUESTIONS IN
MATHEMATICS.

ARMY PRELIMINARY EXAMINATION, 1882-1889, Specimens of Papers set at the. With Answers to the Mathematical Questions. Subjects: Arithmetic, Algebra, Euclid, Geometrical Drawing, Geography, French, English Dictation. Cr. 8vo. 3s. 6d.

CAMBRIDGE SENATE-HOUSE PROBLEMS AND RIDERS, WITH SOLUTIONS:—

1875—PROBLEMS AND RIDERS. By A. G. GREENHILL, F.R.S. Cr. 8vo. 8s. 6d.

1878—SOLUTIONS OF SENATE-HOUSE PROBLEMS. By the Mathematical Moderators and Examiners. Edited by J. W. L. GLAISHER, F.R.S., Fellow of Trinity College, Cambridge. 12s.

CHRISTIE.—A COLLECTION OF ELEMENTARY TEST-QUESTIONS IN PURE AND MIXED MATHEMATICS; with Answers and Appendices on Synthetic Division, and on the Solution of Numerical Equations by Horner's Method. By JAMES R. CHRISTIE, F.R.S. Cr. 8vo. 8s. 6d.

MILNE.—Works by Rev. JOHN J. MILNE, Private Tutor.

WEEKLY PROBLEM PAPERS. With Notes intended for the use of Students preparing for Mathematical Scholarships, and for Junior Members of the Universities who are reading for Mathematical Honours. Pott 8vo. 4s. 6d.

SOLUTIONS TO WEEKLY PROBLEM PAPERS. Cr. 8vo. 10s. 6d.

COMPANION TO WEEKLY PROBLEM PAPERS. Cr. 8vo. 10s. 6d.

SANDHURST MATHEMATICAL PAPERS, for admission into the Royal Military College, 1881-1889. Edited by E. J. BROOKSMITH, B.A., Instructor in Mathematics at the Royal Military Academy, Woolwich. Cr. 8vo.   [*In the Press.*

WOOLWICH MATHEMATICAL PAPERS, for Admission into the Royal Military Academy, Woolwich, 1880-1888 inclusive. Edited by E. J. BROOKSMITH, B.A. Cr. 8vo. 6s.

WOLSTENHOLME.—Works by JOSEPH WOLSTENHOLME, D.Sc., late Professor of Mathematics in the Royal Engineering Coll. Cooper's Hill.

MATHEMATICAL PROBLEMS, on Subjects included in the First and Second Divisions of the Schedule of Subjects for the Cambridge Mathematical Tripos Examination. New Ed., greatly enlarged. 8vo. 18s.

EXAMPLES FOR PRACTICE IN THE USE OF SEVEN-FIGURE LOGARITHMS. 8vo. 5s.

HIGHER PURE MATHEMATICS.

AIRY.—Works by Sir G. B. AIRY, K.C.B., formerly Astronomer-Royal.

ELEMENTARY TREATISE ON PARTIAL DIFFERENTIAL EQUATIONS. With Diagrams. 2d Ed. Cr. 8vo. 5s. 6d.

ON THE ALGEBRAICAL AND NUMERICAL THEORY OF ERRORS OF OBSERVATIONS AND THE COMBINATION OF OBSERVATIONS. 2d. Ed., revised. Cr. 8vo. 6s. 6d.

BOOLE.—THE CALCULUS OF FINITE DIFFERENCES. By G. BOOLE. 3d Ed., revised by J. F. MOULTON, Q.C. Cr. 8vo. 10s. 6d.

CARLL—A TREATISE ON THE CALCULUS OF VARIATIONS. By LEWIS B. CARLL. Arranged with the purpose of Introducing, as well as Illustrating, its Principles to the Reader by means of Problems, and Designed to present in all Important Particulars a Complete View of ·the Present State of the Science. 8vo. 21s.

EDWARDS.—THE DIFFERENTIAL CALCULUS. By JOSEPH EDWARDS, M.A., With Applications and numerous Examples. Cr. 8vo. 10s. 6d.

FERRERS.—AN ELEMENTARY TREATISE ON SPHERICAL HARMONICS, AND SUBJECTS CONNECTED WITH THEM. By Rev. N. M. FERRERS, D.D., F.R.S., Master of Gonville and Caius College, Cambridge. Cr. 8vo. 7s. 6d.

FORSYTH.—A TREATISE ON DIFFERENTIAL EQUATIONS. By ANDREW RUSSELL FORSYTH, F.R.S., Fellow and Assistant Tutor of Trinity College, Cambridge. 2d Ed. 8vo. 14s.

FROST.—AN ELEMENTARY TREATISE ON CURVE TRACING. By PERCIVAL FROST, M.A., D.Sc. 8vo. 12s.

GREENHILL.—DIFFERENTIAL AND INTEGRAL CALCULUS. By A. G. GREENHILL, Professor of Mathematics to the Senior Class of Artillery Officers, Woolwich. Cr. 8vo. 7s. 6d.

JOHNSON.—Works by WILLIAM WOOLSEY JOHNSON, Professor of Mathematics at the U.S. Naval Academy, Annapolis, Maryland.
INTEGRAL CALCULUS, an Elementary Treatise on the; Founded on the Method of Rates or Fluxions. 8vo. 9s.
CURVE TRACING IN CARTESIAN CO-ORDINATES. Cr. 8vo. 4s. 6d.
A TREATISE ON ORDINARY AND DIFFERENTIAL EQUATIONS. Ex. cr. 8vo. 15s.

KELLAND AND TAIT.—INTRODUCTION TO QUATERNIONS, with numerous examples. By P. KELLAND and P. G. TAIT, Professors in the Department of Mathematics in the University of Edinburgh. 2d Ed. Cr. 8vo. 7s. 6d.

KEMPE.—HOW TO DRAW A STRAIGHT LINE: a Lecture on Linkages. By A. B. KEMPE. Illustrated. Cr. 8vo. 1s. 6d.

KNOX.—DIFFERENTIAL CALCULUS FOR BEGINNERS. By ALEXANDER KNOX. Fcap. 8vo. 3s. 6d.

MERRIMAN.—A TEXT-BOOK OF THE METHOD OF LEAST SQUARES. By MANSFIELD MERRIMAN, Professor of Civil Engineering at Lehigh University, U.S.A.' 8vo. 8s. 6d.

MUIR.—Works by THOS. MUIR, Mathematical Master in the High School of Glasgow.
A TREATISE ON THE THEORY OF DETERMINANTS. With graduated sets of Examples. Cr. 8vo. 7s. 6d.
THE THEORY OF DETERMINANTS IN THE HISTORICAL ORDER OF ITS DEVELOPMENT. Part I. Determinants in General. Leibnitz (1693) to Cayley (1841). 8vo. 10s. 6d.

RICE AND JOHNSON.—DIFFERENTIAL CALCULUS, an Elementary Treatise on the; Founded on the Method of Rates or Fluxions. By J. M. RICE, Professor of Mathematics in the United States Navy, and W. W. JOHNSON, Professor of Mathematics at the United States Naval Academy. 3d Ed., revised and corrected. 8vo. 18s. Abridged Ed. 9s.

TODHUNTER.—Works by ISAAC TODHUNTER, F.R.S.
AN ELEMENTARY TREATISE ON THE THEORY OF EQUATIONS. Cr. 8vo. 7s. 6d
A TREATISE ON THE DIFFERENTIAL CALCULUS. Cr. 8vo. 10s. 6d. KEY. Cr. 8vo. 10s. 6d.
A TREATISE ON THE INTEGRAL CALCULUS AND ITS APPLICATIONS. Cr. 8vo. 10s. 6d. KEY. Cr. 8vo. 10s. 6d.
A HISTORY OF THE MATHEMATICAL THEORY OF PROBABILITY, from the time of Pascal to that of Laplace. 8vo. 18s.
AN ELEMENTARY TREATISE ON LAPLACE'S, LAME'S, AND BESSEL'S FUNCTIONS. Cr. 8vo. 10s. 6d.

MECHANICS: Statics, Dynamics, Hydrostatics, Hydrodynamics. (See also Physics.)

ALEXANDER AND THOMSON.—ELEMENTARY APPLIED MECHANICS. By Prof. T. ALEXANDER, and A. W. THOMSON. Part II. Transverse Stress. Cr. 8vo. 10s. 6d.

BALL.—EXPERIMENTAL MECHANICS. A Course of Lectures delivered at the Royal College of Science for Ireland. By Sir R. S. BALL, F.R.S. 2d Ed. Illustrated. Cr. 8vo. 6s.

CHISHOLM.—THE SCIENCE OF WEIGHING AND MEASURING, AND THE STANDARDS OF MEASURE AND WEIGHT. By H. W. CHISHOLM, Warden of the Standards. Illustrated. Cr. 8vo. 4s. 6d.

CLARKE.—A TABLE OF SPECIFIC GRAVITY FOR SOLIDS AND LIQUIDS. (Constants of Nature : Part I.) New Ed., revised and enlarged. By F. W. CLARKE, Chief Chemist, U.S. Geological Survey. 8vo. 12s. 6d. (Published for the Smithsonian Institution, Washington, U.S.A.)

CLIFFORD.—THE ELEMENTS OF DYNAMIC. An Introduction to the Study of Motion and Rest in Solid and Fluid Bodies. By W. K. CLIFFORD. Part I.—Kinematic. Cr. 8vo. Books I—III. 7s. 6d. ; Book IV. and Appendix, 6s.

COTTERILL.—APPLIED MECHANICS : an Elementary General Introduction to the Theory of Structures and Machines. By J. H. COTTERILL, F.R.S., Professor of Applied Mechanics in the Royal Naval College, Greenwich. 8vo. 18s.

COTTERILL AND SLADE.—ELEMENTARY MANUAL OF APPLIED MECHAN-ICS. By Prof. J. H. COTTERILL and J. H. SLADE. Cr. 8vo. [In the Press.

DYNAMICS, SYLLABUS OF ELEMENTARY. Part I. Linear Dynamics. With an Appendix on the Meanings of the Symbols in Physical Equations. Prepared by the Association for the Improvement of Geometrical Teaching. 4to. 1s.

GANGUILLET AND KUTTER.—A GENERAL FORMULA FOR THE UNIFORM FLOW OF WATER IN RIVERS AND OTHER CHANNELS. By E. GAN-GUILLET and W. R. KUTTER, Engineers in Berne, Switzerland. Translated from the German, with numerous Additions, including Tables and Diagrams, and the Elements of over 1200 Gaugings of Rivers, Small Channels, and Pipes in English Measure, by RUDOLPH HERING, Assoc. Am. Soc. C.E., M. Inst. C.E., and JOHN C. TRAUTWINE Jun., Assoc. Am. Soc. C.E., Assoc. Inst. C.E. 8vo. 17s.

GREAVES.—Works by JOHN GREAVES, M.A., Fellow and Mathematical Lecturer at Christ's College, Cambridge.
STATICS FOR BEGINNERS. Gl. 8vo. 3s. 6d.
A TREATISE ON ELEMENTARY STATICS. 2d Ed. Cr. 8vo. 6s. 6d.

HICKS.—ELEMENTARY DYNAMICS OF PARTICLES AND SOLIDS. By W. M. HICKS, Principal and Professor of Mathematics and Physics, Firth College, Sheffield. Cr. 8vo. 6s. 6d.

JELLETT.—A TREATISE ON THE THEORY OF FRICTION. By JOHN H. JELLETT, B.D., late Provost of Trinity College, Dublin. 8vo. 8s. 6d.

KENNEDY.—THE MECHANICS OF MACHINERY. By A. B. W. KENNEDY, F.R.S. Illustrated. Cr. 8vo. 12s. 6d.

LOCK.—Works by Rev. J. B. LOCK, M.A.
ELEMENTARY STATICS. 2d Ed. Gl. 8vo. 4s. 6d.
DYNAMICS FOR BEGINNERS. 3d Ed. Gl. 8vo. 4s. 6d.

MACGREGOR.—KINEMATICS AND DYNAMICS. An Elementary Treatise. By J. G. MACGREGOR, D.Sc., Munro Professor of Physics in Dalhousie College, Halifax, Nova Scotia. Illustrated. Cr. 8vo. 10s. 6d.

PARKINSON.—AN ELEMENTARY TREATISE ON MECHANICS. By S. PARKINSON, D.D., F.R.S., late Tutor and Prælector of St. John's College, Cambridge. 6th Ed., revised. Cr. 8vo. 9s. 6d.

PIRIE.—LESSONS ON RIGID DYNAMICS. By Rev. G. PIRIE, M.A., Professor of Mathematics in the University of Aberdeen. Cr. 8vo. 6s.

REULEAUX.—THE KINEMATICS OF MACHINERY. Outlines of a Theory of Machines. By Prof. F. REULEAUX. Translated and Edited by Prof. A. B. W. KENNEDY, F.R.S. Illustrated. 8vo. 21s.

ROUTH.—Works by EDWARD JOHN ROUTH, D.Sc., LL.D., F.R.S., Hon. Fellow of St. Peter's College, Cambridge.
A TREATISE ON THE DYNAMICS OF THE SYSTEM OF RIGID BODIES. With numerous Examples. Fourth and enlarged Edition. Two Vols. 8vo. Vol. I.—Elementary Parts. 14s. Vol. II.—The Advanced Parts. 14s.
STABILITY OF A GIVEN STATE OF MOTION, PARTICULARLY STEADY MOTION. Adams Prize Essay for 1877. 8vo. 8s. 6d.

SANDERSON.—HYDROSTATICS FOR BEGINNERS. By F. W. SANDERSON, M.A., Assistant Master at Dulwich College. Gl. 8vo. 4s. 6d.

TAIT AND STEELE.—A TREATISE ON DYNAMICS OF A PARTICLE. By Professor TAIT, M.A., and W. J. STEELE, B.A. 6th Ed., revised. Cr. 8vo. 12s.

TODHUNTER.—Works by ISAAC TODHUNTER, F.R.S.
MECHANICS FOR BEGINNERS. 18mo. 4s. 6d. KEY. Cr. 8vo. 6s. 6d.
A TREATISE ON ANALYTICAL STATICS. 5th Ed. Edited by Prof. J. D. EVERETT, F.R.S. Cr. 8vo. 10s. 6d.

PHYSICS : Sound, Light, Heat, Electricity, Elasticity, Attractions, etc. (See also Mechanics.)

AIRY.—Works by Sir G. B. AIRY, K.C.B., formerly Astronomer-Royal.
ON SOUND AND ATMOSPHERIC VIBRATIONS. With the Mathematical Elements of Music. 2d Ed., revised and enlarged. Cr. 8vo. 9s.
A TREATISE ON MAGNETISM. Cr. 8vo. 9s. 6d.
GRAVITATION : an Elementary Explanation of the Principal Perturbations in the Solar System. 2d Ed. Cr. 8vo. 7s. 6d.

CLAUSIUS.—MECHANICAL THEORY OF HEAT. By R. CLAUSIUS. Translated by W. R. BROWNE, M.A. Cr. 8vo. 10s. 6d.

CUMMING.—AN INTRODUCTION TO THE THEORY OF ELECTRICITY. By LINNÆUS CUMMING, M.A., Assistant Master at Rugby. Illustrated. Cr. 8vo. 8s. 6d.

DANIELL.—A TEXT-BOOK OF THE PRINCIPLES OF PHYSICS. By ALFRED DANIELL, D.Sc. Illustrated. 2d Ed., revised and enlarged. 8vo. 21s.

DAY.—ELECTRIC LIGHT ARITHMETIC. By R. E. DAY, Evening Lecturer in Experimental Physics at King's College, London. Pott 8vo. 2s.

EVERETT.—UNITS AND PHYSICAL CONSTANTS. By J. D. EVERETT, F.R.S., Professor of Natural Philosophy, Queen's College, Belfast. 2d Ed. Ex. fcap. 8vo. 5s.

FERRERS.—AN ELEMENTARY TREATISE ON SPHERICAL HARMONICS, and Subjects connected with them. By Rev. N. M. FERRERS, D.D., F.R.S., Master of Gonville and Caius College, Cambridge. Cr. 8vo. 7s. 6d.

FESSENDEN.—A SCHOOL CLASS-BOOK OF PHYSICS. By C. FESSENDEN. Illustrated. Fcp. 8vo. [In the Press.

GRAY.—THE THEORY AND PRACTICE OF ABSOLUTE MEASUREMENTS IN ELECTRICITY AND MAGNETISM. By A. GRAY, F.R.S.E., Professor of Physics in the University College of North Wales. Two Vols. Cr. 8vo. Vol. I. 12s. 6d. [Vol. II. In the Press.
ABSOLUTE MEASUREMENTS IN ELECTRICITY AND MAGNETISM. 2d Ed., revised and greatly enlarged. Fcap. 8vo. 5s. 6d.

IBBETSON.—THE MATHEMATICAL THEORY OF PERFECTLY ELASTIC SOLIDS, with a Short Account of Viscous Fluids. By W. J. IBBETSON, late Senior Scholar of Clare College, Cambridge. 8vo. 21s.

JONES.—EXAMPLES IN PHYSICS. By D. E. JONES, B.Sc., Professor of Physics in the University College of Wales, Aberystwyth. Fcap. 8vo. 3s. 6d.
SOUND, LIGHT, AND HEAT. An Elementary Text-Book. With Illustrations. Fcap. 8vo. [In the Press.

LODGE.—MODERN VIEWS OF ELECTRICITY. By OLIVER J. LODGE, F.R.S., Professor of Experimental Physics in University College, Liverpool. Illustrated. Cr. 8vo. 6s. 6d.

LOEWY.—Works by B. LOEWY, Examiner in Experimental Physics to the College of Preceptors.
QUESTIONS AND EXAMPLES ON EXPERIMENTAL PHYSICS: Sound, Light, Heat, Electricity, and Magnetism. Fcap. 8vo. 2s.
A GRADUATED COURSE OF NATURAL SCIENCE FOR ELEMENTARY AND TECHNICAL SCHOOLS AND COLLEGES. In Three Parts. Part I. FIRST YEAR's COURSE. Gl. 8vo. 2s.

LUPTON.—NUMERICAL TABLES AND CONSTANTS IN ELEMENTARY SCIENCE. By S. LUPTON, M.A., late Assistant Master at Harrow. Ex. fcap. 8vo. 2s. 6d.

MACFARLANE.—PHYSICAL ARITHMETIC. By A. MACFARLANE, D.Sc., late Examiner in Mathematics at the University of Edinburgh. Cr. 8vo. 7s. 6d.

MAYER.—SOUND: a Series of Simple, Entertaining, and Inexpensive Experiments in the Phenomena of Sound. By A. M. MAYER, Professor of Physics in the Stevens Institute of Technology, Illustrated. Cr. 8vo. 3s. 6d.

MAYER AND BARNARD.—LIGHT: a Series of Simple, Entertaining, and Inexpensive Experiments in the Phenomena of Light. By A. M. MAYER and C. BARNARD. Illustrated. Cr. 8vo. 2s. 6d.

MOLLOY.—GLEANINGS IN SCIENCE: Popular Lectures on Scientific Subjects. By the Rev. GERALD MOLLOY, D.Sc., Rector of the Catholic University of Ireland. 8vo. 7s. 6d.

NEWTON.—PRINCIPIA. Edited by Prof. Sir W. THOMSON and Prof. BLACKBURNE. 4to. 31s. 6d.

THE FIRST THREE SECTIONS OF NEWTON'S PRINCIPIA. With Notes and Illustrations. Also a Collection of Problems, principally intended as Examples of Newton's Methods. By P. FROST, M.A., D.Sc. 3d. Ed. 8vo. 12s.

PARKINSON.—A TREATISE ON OPTICS. By S. PARKINSON, D.D., F.R.S., late Tutor and Prælector of St. John's College, Cambridge. 4th Ed., revised and enlarged. Cr. 8vo. 10s. 6d.

PEABODY.—THERMODYNAMICS OF THE STEAM-ENGINE AND OTHER HEAT-ENGINES. By CECIL H. PEABODY, Associate Professor of Steam Engineering, Massachusetts Institute of Technology. 8vo. 21s.

PERRY.— STEAM: an Elementary Treatise. By JOHN PERRY, Professor of Mechanical Engineering and Applied Mechanics at the Technical College, Finsbury. 18mo. 4s. 6d.

PRESTON.—A TREATISE ON THE THEORY OF LIGHT. By THOMAS PRESTON, M.A. Illustrated. 8vo.   [In the Press.

RAYLEIGH.—THE THEORY OF SOUND. By Lord Rayleigh, F.R.S. 8vo. Vol. I. 12s. 6d. Vol. II. 12s. 6d.   [Vol. III. In the Press.

SHANN.—AN ELEMENTARY TREATISE ON HEAT, IN RELATION TO STEAM AND THE STEAM-ENGINE. By G. SHANN, M.A. Illustrated. Cr. 8vo. 4s. 6d.

SPOTTISWOODE.—POLARISATION OF LIGHT. By the late W. SPOTTISWOODE, F.R.S. Illustrated. Cr. 8vo. 3s. 6d.

STEWART.—Works by BALFOUR STEWART, F.R.S., late Langworthy Professor of Physics in the Owens College, Victoria University, Manchester.

PRIMER OF PHYSICS. Illustrated. With Questions. 18mo. 1s.

LESSONS IN ELEMENTARY PHYSICS. Illustrated. Fcap. 8vo. 4s. 6d.

QUESTIONS. By Prof. T. H. CORE. Fcap. 8vo. 2s.

STEWART AND GEE.—LESSONS IN ELEMENTARY PRACTICAL PHYSICS. By BALFOUR STEWART, F.R.S., and W. W. HALDANE GEE, B.Sc. Cr. 8vo. Vol. I. GENERAL PHYSICAL PROCESSES. 6s. Vol. II. ELECTRICITY AND MAGNETISM. 7s. 6d.   [Vol. III. OPTICS, HEAT, and SOUND. In the Press.

PRACTICAL PHYSICS FOR SCHOOLS AND THE JUNIOR STUDENTS OF COLLEGES. Gl. 8vo. Vol. I. ELECTRICITY AND MAGNETISM. 2s. 6d. [Vol. II. OPTICS, HEAT, AND SOUND. In the Press.

STOKES.—ON LIGHT. Burnett Lectures, delivered in Aberdeen in 1883-4-5. By Sir G. G. STOKES, F.R.S., Lucasian Professor of Mathematics in the University of Cambridge. First Course: ON THE NATURE OF LIGHT. Second Course: ON LIGHT AS A MEANS OF INVESTIGATION. Third Course: ON THE BENEFICIAL EFFECTS OF LIGHT. Cr. 8vo. 7s. 6d.

** The 2d and 3d Courses may be had separately. Cr. 8vo. 2s. 6d. each.

STONE.—AN ELEMENTARY TREATISE ON SOUND. By W. H. STONE. Illustrated. Fcap. 8vo. 3s. 6d.

TAIT.—HEAT. By P. G. TAIT, Professor of Natural Philosophy in the University of Edinburgh. Cr. 8vo. 6s.

TAYLOR.—SOUND AND MUSIC. An Elementary Treatise on the Physical Constitution of Musical Sounds and Harmony, including the Chief Acoustical Discoveries of Professor Helmholtz. By SEDLEY TAYLOR, M.A. Illustrated. 2d Ed. Ex. Cr. 8vo. 8s. 6d.

THOMPSON. — ELEMENTARY LESSONS IN ELECTRICITY AND MAGNET-ISM. By SILVANUS P. THOMPSON, Principal and Professor of Physics in the Technical College, Finsbury. Illustrated. New Ed., revised. Fcap. 8vo. 4s. 6d.

THOMSON.—Works by J. J. THOMSON, Professor of Experimental Physics in the University of Cambridge.
A TREATISE ON THE MOTION OF VORTEX RINGS. Adams Prize Essay, 1882. 8vo. 6s.
APPLICATIONS OF DYNAMICS TO PHYSICS AND CHEMISTRY. Cr. 8vo. 7s. 6d.

THOMSON.—Works by Sir W. THOMSON, F.R.S., Professor of Natural Philosophy in the University of Glasgow.
ELECTROSTATICS AND MAGNETISM, REPRINTS OF PAPERS ON. 2d Ed. 8vo. 18s.
POPULAR LECTURES AND ADDRESSES. 3 Vols. Illustrated. Cr. 8vo. Vol. I. CONSTITUTION OF MATTER. 6s.

TODHUNTER.—Works by ISAAC TODHUNTER, F.R.S.
NATURAL PHILOSOPHY FOR BEGINNERS. Part I. The properties of Solid and Fluid Bodies. 18mo. 3s. 6d. Part II. Sound, Light, and Heat. 18mo. 3s. 6d.
AN ELEMENTARY TREATISE ON LAPLACE'S, LAME'S, AND BESSEL'S FUNCTIONS. Crown 8vo. 10s. 6d.
A HISTORY OF THE MATHEMATICAL THEORIES OF ATTRACTION, AND THE FIGURE OF THE EARTH, from the time of Newton to that of Laplace. 2 vols. 8vo. 24s.

TURNER.—A COLLECTION OF EXAMPLES ON HEAT AND ELECTRICITY. By H. H. TURNER, Fellow of Trinity College, Cambridge. Cr. 8vo. 2s. 6d.

WRIGHT.—LIGHT: A Course of Experimental Optics, chiefly with the Lantern. By LEWIS WRIGHT. Illustrated. Cr. 8vo. 7s. 6d.

ASTRONOMY.

AIRY.—Works by Sir G. B. AIRY, K.C.B., formerly Astronomer-Royal.
POPULAR ASTRONOMY. 18mo. 4s. 6d.
GRAVITATION: an Elementary Explanation of the Principal Perturbations in the Solar System. 2d Ed. Cr. 8vo. 7s. 6d.

CHEYNE.—AN ELEMENTARY TREATISE ON THE PLANETARY THEORY. By C. H. H CHEYNE. With Problems. 3d Ed. Edited by Rev. A. FREEMAN, M.A., F.R.A.S. Cr. 8vo. 7s. 6d.

FORBES.—TRANSIT OF VENUS. By G. FORBES, Professor of Natural Philosophy in the Andersonian University, Glasgow. Illustrated. Cr. 8vo. 3s. 6d.

GODFRAY.—Works by HUGH GODFRAY, M.A., Mathematical Lecturer at Pembroke College, Cambridge.
A TREATISE ON ASTRONOMY. 4th Ed. 8vo. 12s. 6d.
AN ELEMENTARY TREATISE ON THE LUNAR THEORY, with a brief Sketch of the Problem up to the time of Newton. 2d Ed., revised. Cr. 8vo. 5s. 6d.

LOCKYER.—Works by J. NORMAN LOCKYER, F.R.S.
PRIMER OF ASTRONOMY. Illustrated. 18mo. 1s.
ELEMENTARY LESSONS IN ASTRONOMY.—With Spectra of the Sun, Stars, and Nebulæ, and numerous Illustrations. 36th Thousand. Revised throughout. Fcap. 8vo. 5s. 6d.
QUESTIONS ON LOCKYER'S ELEMENTARY LESSONS IN ASTRONOMY. By J. FORBES ROBERTSON. 18mo. 1s. 6d.
THE CHEMISTRY OF THE SUN. Illustrated. 8vo. 14s.
THE METEORIC HYPOTHESIS. Illustrated. 8vo. [In the Press.
THE EVOLUTION OF THE HEAVENS AND THE EARTH. Cr. 8vo. Illustrated. [In the Press.

NEWCOMB.—POPULAR ASTRONOMY. By S. NEWCOMB, LL.D., Professor U.S. Naval Observatory. Illustrated. 2d Ed., revised. 8vo. 18s.

HISTORICAL.

BALL.—A SHORT ACCOUNT OF THE HISTORY OF MATHEMATICS. By W.
W. R. BALL. Cr. 8vo. 10s. 6d.

TODHUNTER.—Works by ISAAC TODHUNTER, F.R.S.
A HISTORY OF THE MATHEMATICAL THEORY OF PROBABILITY from
the time of Pascal to that of Laplace. 8vo. 18s.
A HISTORY OF THE MATHEMATICAL THEORIES OF ATTRACTION,
AND THE FIGURE OF THE EARTH, from the time of Newton to that of
Laplace. 2 vols. 8vo. 24s.

NATURAL SCIENCES.

Chemistry; Physical Geography, Geology, and Mineralogy; Biology;
Medicine.

(FOR MECHANICS, PHYSICS, AND ASTRONOMY, see
MATHEMATICS.)

CHEMISTRY.

ARMSTRONG.—A MANUAL OF INORGANIC CHEMISTRY. By HENRY ARM-
STRONG, F.R.S., Professor of Chemistry in the City and Guilds of London Technical Institute. Cr. 8vo.   [*In preparation.*

COHEN.—THE OWENS COLLEGE COURSE OF PRACTICAL ORGANIC
CHEMISTRY. By JULIUS B. COHEN, Ph.D., Assistant Lecturer on Chemistry
in the Owens College, Manchester. With a Preface by Sir HENRY ROSCOE,
F.R.S., and C. SCHORLEMMER, F.R.S. Fcap. 8vo. 2s. 6d.

COOKE.—ELEMENTS OF CHEMICAL PHYSICS. By JOSIAH P. COOKE, Jun.
Erving Professor of Chemistry and Mineralogy in Harvard University. 4th Ed.
8vo. 21s.

FLEISCHER.—A SYSTEM OF VOLUMETRIC ANALYSIS. By EMIL FLEISCHER.
Translated, with Notes and Additions, by M. M. P. MUIR, F.R.S.E. Illustrated.
Cr. 8vo. 7s. 6d.

FRANKLAND.—A HANDBOOK OF AGRICULTURAL CHEMICAL ANALYSIS.
By P. F. FRANKLAND, F.R.S., Professor of Chemistry in University College,
Dundee. Cr. 8vo. 7s. 6d.

HARTLEY.—A COURSE OF QUANTITATIVE ANALYSIS FOR STUDENTS.
By W. NOEL HARTLEY, F.R.S., Professor of Chemistry and of Applied Chemistry, Science and Art Department, Royal College of Science, Dublin. Gl.
8vo. 5s.

HIORNS.—PRACTICAL METALLURGY AND ASSAYING. A Text-book for
the use of Teachers, Students, and Assayers. By ARTHUR H. HIORNS, Prin-
cipal of the School of Metallurgy, Birmingham and Midland Institute. Illus-
trated. Gl. 8vo. 6s.
A TEXT-BOOK OF ELEMENTARY METALLURGY FOR THE USE OF
STUDENTS. To which is added an Appendix of Examination Questions, em-
bracing the whole of the Questions set in the three stages of the subject by the
Science and Art Department for the past twenty years. By the Same. Gl. 8vo. 4s.
IRON AND STEEL MANUFACTURE. A Text-Book for Beginners. By the
Same. Illustrated. Gl. 8vo. 3s. 6d.
MIXED METALS AND METALLIC ALLOYS. By the Same. [*In the Press.*

JONES.—THE OWENS COLLEGE JUNIOR COURSE OF PRACTICAL CHEM-
ISTRY. By FRANCIS JONES, F.R.S.E., Chemical Master at the Grammar School,
Manchester. With Preface by Sir HENRY ROSCOE, F.R.S. Illustrated. Fcp.
8vo. 2s. 6d.

THORPE AND RÜCKER.—A TREATISE ON CHEMICAL PHYSICS. By Prof. T. E. THORPE, F.R.S., and Prof. A. W. RÜCKER, F.R.S. Illustrated. 8vo.
[In preparation.

WRIGHT.—METALS AND THEIR CHIEF INDUSTRIAL APPLICATIONS. By C. ALDER WRIGHT, Lecturer on Chemistry in St. Mary's Hospital School. Ex. fcap. 8vo. 3s. 6d.

PHYSICAL GEOGRAPHY, GEOLOGY, AND MINERALOGY.

BLANFORD.—THE RUDIMENTS OF PHYSICAL GEOGRAPHY FOR THE USE OF INDIAN SCHOOLS; with a Glossary of Technical Terms employed. By H. F. BLANFORD, F.G.S. Illustrated. Cr. 8vo. 2s. 6d.

FERREL.—A POPULAR TREATISE ON THE WINDS. Comprising the General Motions of the Atmosphere, Monsoons, Cyclones, Tornadoes, Waterspouts, Hailstorms, &c. By WILLIAM FERREL, M.A., Member of the American National Academy of Sciences. 8vo. 18s.

FISHER.—PHYSICS OF THE EARTH'S CRUST. By the Rev. OSMOND FISHER, M.A., F.G.S., Hon. Fellow of King's College, London. 2d Ed., altered and enlarged. 8vo. 12s.

GEIKIE.—Works by ARCHIBALD GEIKIE, LL.D., F.R.S., Director-General of the Geological Survey of Great Britain and Ireland.
PRIMER OF PHYSICAL GEOGRAPHY. Illustrated. With Questions. 18mo. 1s.
ELEMENTARY LESSONS IN PHYSICAL GEOGRAPHY. Illustrated. Fcap. 8vo. 4s. 6d. QUESTIONS ON THE SAME. 1s. 6d.
PRIMER OF GEOLOGY. Illustrated. 18mo. 1s.
CLASS BOOK OF GEOLOGY. Illustrated. New and Cheaper Edition.
TEXT-BOOK OF GEOLOGY. Illustrated. 2d Ed., 7th Thousand, revised and enlarged. 8vo. 28s.
OUTLINES OF FIELD GEOLOGY. Illustrated. Ex. fcap. 8vo. 3s. 6d.
THE SCENERY AND GEOLOGY OF SCOTLAND, VIEWED IN CONNEXION WITH ITS PHYSICAL GEOLOGY. Illustrated. Cr. 8vo. 12s. 6d.

HUXLEY.—PHYSIOGRAPHY. An Introduction to the Study of Nature. By T. H. HUXLEY, F.R.S. Illustrated. New and Cheaper Edition. Cr. 8vo. 6s.

LOCKYER.—OUTLINES OF PHYSIOGRAPHY—THE MOVEMENTS OF THE EARTH. By J. NORMAN LOCKYER, F.R.S., Examiner in Physiography for the Science and Art Department. Illustrated. Cr. 8vo. Sewed, 1s. 6d.

PHILLIPS.—A TREATISE ON ORE DEPOSITS. By J. ARTHUR PHILLIPS, F.R.S. Illustrated. 8vo. 25s.

ROSENBUSCH AND IDDINGS.— MICROSCOPICAL PHYSIOGRAPHY OF THE ROCK-MAKING MINERALS: AN AID TO THE MICROSCOPICAL STUDY OF ROCKS. By H. ROSENBUSCH. Translated and Abridged by J. P. IDDINGS. Illustrated. 8vo. 24s.

BIOLOGY.

ALLEN.—ON THE COLOURS OF FLOWERS, as Illustrated in the British Flora. By GRANT ALLEN. Illustrated. Cr. 8vo. 3s. 6d.

BALFOUR.—A TREATISE ON COMPARATIVE EMBRYOLOGY. By F. M. BALFOUR, F.R.S., Fellow and Lecturer of Trinity College, Cambridge. Illustrated. 2d Ed., reprinted without alteration from the 1st Ed. 2 vols. 8vo. Vol. I. 18s. Vol. II. 21s.

BALFOUR AND WARD.—A GENERAL TEXT-BOOK OF BOTANY. By ISAAC BAYLEY BALFOUR, F.R.S., Professor of Botany in the University of Edinburgh, and H. MARSHALL WARD, F.R.S., Professor of Botany in the Royal Indian Engineering College, Cooper's Hill. 8vo. *[In preparation.*

BETTANY.—FIRST LESSONS IN PRACTICAL BOTANY. By G. T. BETTANY. 18mo. 1s.

BOWER.—A COURSE OF PRACTICAL INSTRUCTION IN BOTANY. By F. O. BOWER, D.Sc., Regius Professor of Botany in the University of Glasgow. Cr. 8vo. 10s. 6d.

CHURCH AND SCOTT.—MANUAL OF VEGETABLE PHYSIOLOGY. By Professor A. H. CHURCH, and D. H. SCOTT, D.Sc., Lecturer in the Normal School of Science. Illustrated. Cr. 8vo. [In preparation.

COPE.—THE ORIGIN OF THE FITTEST. Essays on Evolution. By E. D. COPE, M.A., Ph.D. 8vo. 12s. 6d.

COUES.—FIELD ORNITHOLOGY AND GENERAL ORNITHOLOGY. By ELLIOTT COUES, M.A. Illustrated. 8vo. [In the Press.

DARWIN.—MEMORIAL NOTICES OF CHARLES DARWIN, F.R.S., &c. By T. H. HUXLEY, F.R.S., G. J. ROMANES, F.R.S., ARCHIBALD GEIKIE, F.R.S., and W. T. THISELTON DYER, F.R.S. Reprinted from Nature. With a Portrait. Cr. 8vo. 2s. 6d.

EIMER.—ORGANIC EVOLUTION AS THE RESULT OF THE INHERITANCE OF ACQUIRED CHARACTERS ACCORDING TO THE LAWS OF ORGANIC GROWTH. By Dr. G. H. THEODOR EIMER. Translated by J. T. CUNNINGHAM, F.R.S.E., late Fellow of University College, Oxford. 8vo. 12s. 6d.

FEARNLEY.—A MANUAL OF ELEMENTARY PRACTICAL HISTOLOGY. By WILLIAM FEARNLEY. Illustrated. Cr. 8vo. 7s. 6d.

FLOWER AND GADOW.—AN INTRODUCTION TO THE OSTEOLOGY OF THE MAMMALIA. By W. H. FLOWER, F.R.S., Director of the Natural History Departments of the British Museum. Illustrated. 3d Ed. Revised with the assistance of HANS GADOW, Ph.D., Lecturer on the Advanced Morphology of Vertebrates in the University of Cambridge. Cr. 8vo. 10s. 6d.

FOSTER.—Works by MICHAEL FOSTER, M.D., Professor of Physiology in the University of Cambridge.

PRIMER OF PHYSIOLOGY. Illustrated. 18mo. 1s.

A TEXT-BOOK OF PHYSIOLOGY. Illustrated. 5th Ed., largely revised. In Three Parts. 8vo. Part I., comprising Book I. Blood—The Tissues of Movement, The Vascular Mechanism. 10s. 6d. Part II., comprising Book II. The Tissues of Chemical Action, with their Respective Mechanisms—Nutrition. 10s. 6d. [Part III. In the Press.

FOSTER AND BALFOUR.—THE ELEMENTS OF EMBRYOLOGY. By Prof. MICHAEL FOSTER, M.D., and the late F. M. BALFOUR, F.R.S., Professor of Animal Morphology in the University of Cambridge. 2d Ed., revised. Edited by A. SEDGWICK, M.A., Fellow and Assistant Lecturer of Trinity College, Cambridge, and W. HEAPE, M.A., late Demonstrator in the Morphological Laboratory of the University of Cambridge. Illustrated. Cr. 8vo. 10s. 6d.

FOSTER AND LANGLEY.—A COURSE OF ELEMENTARY PRACTICAL PHYSIOLOGY AND HISTOLOGY. By Prof. MICHAEL FOSTER, M.D., and J. N. LANGLEY, F.R.S., Fellow of Trinity College, Cambridge. 6th Ed. Cr. 8vo. 7s. 6d.

GAMGEE.—A TEXT-BOOK OF THE PHYSIOLOGICAL CHEMISTRY OF THE ANIMAL BODY. Including an Account of the Chemical Changes occurring in Disease. By A. GAMGEE, M.D., F.R.S. Illustrated. 8vo. Vol. I. 18s.

GOODALE.—PHYSIOLOGICAL BOTANY. I. Outlines of the Histology of Phænogamous Plants. II. Vegetable Physiology. By GEORGE LINCOLN GOODALE, M.A., M.D., Professor of Botany in Harvard University. 8vo. 10s. 6d.

GRAY.—STRUCTURAL BOTANY, OR ORGANOGRAPHY ON THE BASIS OF MORPHOLOGY. To which are added the Principles of Taxonomy and Phytography, and a Glossary of Botanical Terms. By Prof. ASA GRAY, LL.D. 8vo. 10s. 6d.

THE SCIENTIFIC PAPERS OF ASA GRAY. Selected by C. SPRAGUE SARGENT. 2 vols. Vol. I. Reviews of Works on Botany and Related Subjects, 1834-1887. Vol. II. Essays, Biographical Sketches, 1841-1886. 8vo. 21s.

HAMILTON.—A SYSTEMATIC AND PRACTICAL TEXT-BOOK OF PATHOLOGY. By D. J. HAMILTON, F.R.S.E., Professor of Pathological Anatomy in the University of Aberdeen. Illustrated. 8vo. Vol. I. 25s.

HOOKER.—Works by Sir JOSEPH HOOKER, F.R.S., &c.
  PRIMER OF BOTANY. Illustrated. 18mo. 1s.
  THE STUDENT'S FLORA OF THE BRITISH ISLANDS. 3d Ed., revised. Gl. 8vo. 10s. 6d.

HOWES.—AN ATLAS OF PRACTICAL ELEMENTARY BIOLOGY. By G. B. Howes, Assistant Professor of Zoology, Normal School of Science and Royal School of Mines. With a Preface by Prof. T. H. HUXLEY, F.R.S. 4to. 14s.

HUXLEY.—Works by Prof. T. H. HUXLEY, F.R.S.
  INTRODUCTORY PRIMER OF SCIENCE. 18mo. 1s.
  LESSONS IN ELEMENTARY PHYSIOLOGY. Illustrated. Fcap. 8vo. 4s. 6d.
  QUESTIONS ON HUXLEY'S PHYSIOLOGY. By T. ALCOCK, M.D. 18mo. 1s. 6d.

HUXLEY AND MARTIN.—A COURSE OF PRACTICAL INSTRUCTION IN ELEMENTARY BIOLOGY. By Prof. T. H. HUXLEY, F.R.S., assisted by H. N. MARTIN, F.R.S., Professor of Biology in the Johns Hopkins University, U.S.A. New Ed., revised and extended by G. B. HOWES and D. H. SCOTT, Ph.D., Assistant Professors, Normal School of Science and Royal School of Mines. With a Preface by T. H. HUXLEY, F.R.S. Cr. 8vo. 10s. 6d.

KLEIN.—Works by E. KLEIN, F.R.S., Lecturer on General Anatomy and Physiology in the Medical School of St. Bartholomew's Hospital, Professor of Bacteriology at the College of State Medicine, London.
  MICRO-ORGANISMS AND DISEASE. An Introduction into the Study of Specific Micro-Organisms. Illustrated. 3d Ed., revised. Cr. 8vo. 6s.
  THE BACTERIA IN ASIATIC CHOLERA. Cr. 8vo. 5s.

LANG.—TEXT-BOOK OF COMPARATIVE ANATOMY. By Dr. ARNOLD LANG, Professor of Zoology in the University of Zurich, Translated by H. M. BERNARD, M.A., and M. BERNARD. 2 vols. 8vo.   [In the Press.

LANKESTER.—Works by E. RAY LANKESTER, F.R.S., Professor of Zoology in University College, London.
  A TEXT-BOOK OF ZOOLOGY. 8vo.   [In preparation.
  DEGENERATION : A CHAPTER IN DARWINISM. Illustrated. Cr. 8vo. 2s. 6d.
  THE ADVANCEMENT OF SCIENCE. Occasional Essays and Addresses. 8vo. 10s. 6d.

LUBBOCK.—Works by the Right Hon. Sir JOHN LUBBOCK, F.R.S., D.C.L.
  THE ORIGIN AND METAMORPHOSES OF INSECTS. Illustrated. Cr. 8vo. 3s. 6d.
  ON BRITISH WILD FLOWERS CONSIDERED IN RELATION TO INSECTS. Illustrated. Cr. 8vo. 4s. 6d.
  FLOWERS, FRUITS, AND LEAVES. Illustrated. 2d Ed. Cr. 8vo. 4s. 6d.
  SCIENTIFIC LECTURES. 2d Ed. 8vo. 8s. 6d.
  FIFTY YEARS OF SCIENCE. Being the Address delivered at York to the British Association, August 1881. 5th Ed. Cr. 8vo. 2s. 6d.

MARTIN AND MOALE.—ON THE DISSECTION OF VERTEBRATE ANIMALS. By Prof. H. N. MARTIN and W. A. MOALE. Cr. 8vo. [In preparation.

MIVART.—LESSONS IN ELEMENTARY ANATOMY. By ST. GEORGE MIVART, F.R.S., Lecturer on Comparative Anatomy at St. Mary's Hospital. Illustrated. Fcap. 8vo. 6s. 6d.

MULLER.—THE FERTILISATION OF FLOWERS. By HERMANN MÜLLER. Translated and Edited by D'ARCY W. THOMPSON, B.A., Professor of Biology in University College, Dundee. With a Preface by C. DARWIN, F.R.S. Illustrated. 8vo. 21s.

OLIVER.—Works by DANIEL OLIVER, F.R.S., late Professor of Botany in University College, London.
  LESSONS IN ELEMENTARY BOTANY. Illustrated. Fcap. 8vo. 4s. 6d.
  FIRST BOOK OF INDIAN BOTANY. Illustrated. Ex. fcap. 8vo. 6s. 6d.

PARKER.—Works by T. JEFFREY PARKER, F.R.S., Professor of Biology in the University of Otago, New Zealand

A COURSE OF INSTRUCTION IN ZOOTOMY (VERTEBRATA). Illustrated. Cr. 8vo. 8s. 6d.

LESSONS IN ELEMENTARY BIOLOGY. Illustrated. Cr. 8vo. [In the Press.

PARKER AND BETTANY.—THE MORPHOLOGY OF THE SKULL. By Prof. W. K. PARKER, F.R.S., and G. T. BETTANY. Illustrated. Cr. 8vo. 10s. 6d.

ROMANES.—THE SCIENTIFIC EVIDENCES OF ORGANIC EVOLUTION. By GEORGE J. ROMANES, F.R.S., Zoological Secretary of the Linnean Society. Cr. 8vo. 2s. 6d.

SEDGWICK.—A SUPPLEMENT TO F. M. BALFOUR'S TREATISE ON EM-BRYOLOGY. By ADAM SEDGWICK, F.R.S., Fellow and Lecturer of Trinity College, Cambridge. Illustrated. 8vo. [In preparation.

SHUFELDT.—THE MYOLOGY OF THE RAVEN (Corvus corax Sinuatus). A Guide to the Study of the Muscular System in Birds. By R. W. SHUFELDT. Illustrated. 8vo. [In the Press.

SMITH.—DISEASES OF FIELD AND GARDEN CROPS, CHIEFLY SUCH AS ARE CAUSED BY FUNGI. By W. G. SMITH, F.L.S. Illustrated. Fcap. 8vo. 4s. 6d.

STEWART AND CORRY.—A FLORA OF THE NORTH-EAST OF IRELAND. Including the Phanerogamia, the Cryptogamia Vascularia, and the Muscineæ. By S. A. STEWART, Curator of the Collections in the Belfast Museum, and the late T. H. CORRY, M.A., Lecturer on Botany in the University Medical and Science Schools, Cambridge. Cr. 8vo. 5s. 6d.

WALLACE.—DARWINISM: An Exposition of the Theory of Natural Selection, with some of its Applications. By ALFRED RUSSEL WALLACE, LL.D., F.R.S. 3d Ed. Cr. 8vo. 9s.

WARD.—TIMBER AND SOME OF ITS DISEASES. By H. MARSHALL WARD, F.R.S., Professor of Botany in the Royal Indian Engineering College, Cooper's Hill. Illustrated. Cr. 8vo. 6s.

WIEDERSHEIM.—ELEMENTS OF THE COMPARATIVE ANATOMY OF VERTEBRATES. By Prof. R. WIEDERSHEIM. Adapted by W. NEWTON PAR-KER, Professor of Biology in the University College of South Wales and Mon-mouthshire. With Additions. Illustrated. 8vo. 12s. 6d.

MEDICINE.

BLYTH.—A MANUAL OF PUBLIC HEALTH. By A. WYNTER BLYTH, M.R.C.S. 8vo. [In the Press.

BRUNTON.—Works by T. LAUDER BRUNTON, M.D., F.R.S., Examiner in Materia Medica in the University of London, in the Victoria University, and in the Royal College of Physicians, London.

A TEXT-BOOK OF PHARMACOLOGY, THERAPEUTICS, AND MATERIA MEDICA. Adapted to the United States Pharmacopœia by F. H. WILLIAMS, M.D., Boston, Mass. 3d Ed. Adapted to the New British Pharmacopœia, 1885. 8vo. 21s.

TABLES OF MATERIA MEDICA: A Companion to the Materia Medica Museum. Illustrated. Cheaper Issue. 8vo. 5s.

ON THE CONNECTION BETWEEN CHEMICAL CONSTITUTION AND PHYSIOLOGICAL ACTION, BEING AN INTRODUCTION TO MODERN THERAPEUTICS. Croonian Lectures. 8vo. [In the Press.

GRIFFITHS.—LESSONS ON PRESCRIPTIONS AND THE ART OF PRE-SCRIBING. By W. HANDSEL GRIFFITHS. Adapted to the Pharmacopœia, 1885. 18mo. 3s. 6d.

HAMILTON.—A TEXT-BOOK OF PATHOLOGY, SYSTEMATIC AND PRAC-TICAL. By D. J. HAMILTON, F.R.S.E., Professor of Pathological Anatomy, University of Aberdeen. Illustrated. Vol. I. 8vo. 25s.

KLEIN.—Works by E. KLEIN, F.R.S., Lecturer on General Anatomy and Physio-logy in the Medical School of St. Bartholomew's Hospital, London.

MICRO-ORGANISMS AND DISEASE. An Introduction into the Study of Specific Micro-Organisms. Illustrated. 3d Ed., revised. Cr. 8vo. 6s.

THE BACTERIA IN ASIATIC CHOLERA. Cr. 8vo. 5s.

WHITE.—A TEXT-BOOK OF GENERAL THERAPEUTICS. By W. Hale White, M.D., Senior Assistant Physician to and Lecturer in Materia Medica at Guy's Hospital. Illustrated. Cr. 8vo. 8s. 6d.

ZIEGLER—MACALISTER.—TEXT-BOOK OF PATHOLOGICAL ANATOMY AND PATHOGENESIS. By Prof. E. Ziegler. Translated and Edited by Donald Macalister, M.A., M.D., Fellow and Medical Lecturer of St. John's College, Cambridge. Illustrated. 8vo.

    Part I.—GENERAL PATHOLOGICAL ANATOMY. 2d Ed. 12s. 6d.

    Part II.—SPECIAL PATHOLOGICAL ANATOMY. Sections I.-VIII. 2d Ed. 12s. 6d. Sections IX.-XII. 12s. 6d.

HUMAN SCIENCES.

Mental and Moral Philosophy; Political Economy; Law and Politics; Anthropology; Education.

MENTAL AND MORAL PHILOSOPHY.

BOOLE.—THE MATHEMATICAL ANALYSIS OF LOGIC. Being an Essay towards a Calculus of Deductive Reasoning. By George Boole. 8vo. 5s.

CALDERWOOD.—HANDBOOK OF MORAL PHILOSOPHY. By Rev. Henry Calderwood, LL.D., Professor of Moral Philosophy in the University of Edinburgh. 14th Ed., largely rewritten. Cr. 8vo. 6s.

CLIFFORD.—SEEING AND THINKING. By the late Prof. W. K. Clifford, F.R.S. With Diagrams. Cr. 8vo. 3s. 6d.

JARDINE.—THE ELEMENTS OF THE PSYCHOLOGY OF COGNITION. By Rev. Robert Jardine, D.Sc. 3d Ed., revised. Cr. 8vo. 6s. 6d.

JEVONS.—Works by W. Stanley Jevons, F.R.S.

    PRIMER OF LOGIC. 18mo. 1s.

    ELEMENTARY LESSONS IN LOGIC; Deductive and Inductive, with Copious Questions and Examples, and a Vocabulary of Logical Terms. Fcap. 8vo. 3s. 6d.

    THE PRINCIPLES OF SCIENCE. A Treatise on Logic and Scientific Method. New and revised Ed. Cr. 8vo. 12s. 6d.

    STUDIES IN DEDUCTIVE LOGIC. 2d Ed. Cr. 8vo. 6s.

    PURE LOGIC: AND OTHER MINOR WORKS. Edited by R. Adamson, M.A., LL.D., Professor of Logic at Owens College, Manchester, and Harriet A. Jevons. With a Preface by Prof. Adamson. 8vo. 10s. 6d.

KANT—MAX MÜLLER.—CRITIQUE OF PURE REASON. By Immanuel Kant. 2 vols. 8vo. 16s. each. Vol. I. HISTORICAL INTRODUCTION, by Ludwig Noiré; Vol. II. CRITIQUE OF PURE REASON, translated by F. Max Müller.

KANT—MAHAFFY AND BERNARD.—KANT'S CRITICAL PHILOSOPHY FOR ENGLISH READERS. By J. P. Mahaffy, D.D., Professor of Ancient History in the University of Dublin, and John H. Bernard, B.D., Fellow of Trinity College, Dublin. A new and complete Edition in 2 vols. Cr. 8vo.

    Vol. I. The Kritik of Pure Reason Explained and Defended. 7s. 6d.

    Vol. II. The Prolegomena. Translated with Notes and Appendices. 6s.

KEYNES.—FORMAL LOGIC, Studies and Exercises in. Including a Generalisation of Logical Processes in their application to Complex Inferences. By John Neville Keynes, M.A. 2d Ed., revised and enlarged. Cr. 8vo. 10s. 6d.

McCOSH.—Works by James McCosh, D.D., President of Princeton College.

    PSYCHOLOGY. Cr. 8vo.

    I. THE COGNITIVE POWERS. 6s. 6d.

    II. THE MOTIVE POWERS. 6s. 6d.

    FIRST AND FUNDAMENTAL TRUTHS: being a Treatise on Metaphysics. Ex. cr. 8vo. 9s.

MAURICE.—MORAL AND METAPHYSICAL PHILOSOPHY. By F. D. MAURICE, M.A., late Professor of Moral Philosophy in the University of Cambridge. Vol. I.—Ancient Philosophy and the First to the Thirteenth Centuries. Vol. II.—Fourteenth Century and the French Revolution, with a glimpse into the Nineteenth Century. 4th Ed. 2 vols. 8vo. 16s.

RAY.—A TEXT-BOOK OF DEDUCTIVE LOGIC FOR THE USE OF STUDENTS. By P. K. RAY, D.Sc., Professor of Logic and Philosophy, Presidency College, Calcutta. 4th Ed. Globe 8vo. 4s. 6d.

SIDGWICK.—Works by HENRY SIDGWICK, LL.D. D.C.L., Knightbridge Professor of Moral Philosophy in the University of Cambridge.

THE METHODS OF ETHICS. 3d Ed. 8vo. 14s. A Supplement to the 2d Ed., containing all the important Additions and Alterations in the 3d Ed. 8vo. 6s.

OUTLINES OF THE HISTORY OF ETHICS, for English Readers. 2d Ed., revised. Cr. 8vo. 3s. 6d.

VENN.—Works by JOHN VENN, F.R.S., Examiner in Moral Philosophy in the University of London.

THE LOGIC OF CHANCE. An Essay on the Foundations and Province of the Theory of Probability, with special Reference to its Logical Bearings and its Application to Moral and Social Science. 3d Ed., rewritten and greatly enlarged. Cr. 8vo. 10s. 6d

SYMBOLIC LOGIC. Cr. 8vo. 10s. 6d.

THE PRINCIPLES OF EMPIRICAL OR INDUCTIVE LOGIC. 8vo. 18s.

POLITICAL ECONOMY.

BOHM-BAWERK.—CAPITAL AND INTEREST. Translated by WILLIAM SMART, M.A. 8vo. 14s.

CAIRNES.—THE CHARACTER AND LOGICAL METHOD OF POLITICAL ECONOMY. By J. E. CAIRNES. Cr. 8vo. 6s.

SOME LEADING PRINCIPLES OF POLITICAL ECONOMY NEWLY EXPOUNDED. By the Same. 8vo. 14s.

COSSA.—GUIDE TO THE STUDY OF POLITICAL ECONOMY. By Dr. L. COSSA. Translated. With a Preface by W. S. JEVONS, F.R.S. Cr. 8vo. 4s. 6d.

FAWCETT.—POLITICAL ECONOMY FOR BEGINNERS, WITH QUESTIONS. By Mrs. HENRY FAWCETT. 7th Ed. 18mo. 2s. 6d.

TALES IN POLITICAL ECONOMY. By the Same. Cr. 8vo. 3s.

FAWCETT.—A MANUAL OF POLITICAL ECONOMY. By Right. Hon. HENRY FAWCETT, F.R.S. 7th Ed., revised. With a Chapter on "State Socialism and the Nationalisation of the Land," and an Index. Cr. 8vo. 12s. 6d.

AN EXPLANATORY DIGEST of the above. By C. A. WATERS, B.A. Cr. 8vo. 2s. 6d.

GUNTON.—WEALTH AND PROGRESS : A Critical Examination of the Wages Question and its Economic Relation to Social Reform. By GEORGE GUNTON. Cr. 8vo. 6s.

HOWELL.—THE CONFLICTS OF CAPITAL AND LABOUR. Historically and Economically considered, being a History and Review of the Trade Unions of Great Britain, showing their origin, Progress, Constitution, and Objects, in their varied Political, Social, Economical, and Industrial Aspects. By GEORGE HOWELL, M.P. 2d Ed. revised. Cr. 8vo. 6s.

JEVONS.—Works by W. STANLEY JEVONS, F.R.S.

PRIMER OF POLITICAL ECONOMY. 18mo. 1s.

THE THEORY OF POLITICAL ECONOMY. 3d Ed., revised. 8vo. 10s. 6d.

KEYNES.—THE SCOPE AND METHOD OF POLITICAL ECONOMY. By J. N. KEYNES, M.A. [In preparation.

MARSHALL.—THE ECONOMICS OF INDUSTRY. By A. MARSHALL, M.A., Professor of Political Economy in the University of Cambridge, and MARY P. MARSHALL. Ex. fcap. 8vo. 2s. 6d.

MARSHALL.—THE PRINCIPLES OF ECONOMICS. By ALFRED MARSHALL, M.A. 2 vols. 8vo. [Vol. I. Shortly.

PALGRAVE.—A DICTIONARY OF POLITICAL ECONOMY.   By various
Writers. Edited by R. H. INGLIS PALGRAVE.   [*In the Press.*
SIDGWICK.—THE PRINCIPLES OF POLITICAL ECONOMY. By HENRY
SIDGWICK, LL.D., D.C.L., Knightbridge Professor of Moral Philosophy in the
University of Cambridge. 2d Ed., revised. 8vo. 16s.
WALKER.—Works by FRANCIS A. WALKER, M.A.
FIRST LESSONS IN POLITICAL ECONOMY. Cr. 8vo. 5s.
A BRIEF TEXT-BOOK OF POLITICAL ECONOMY. Cr. 8vo. 6s. 6d.
POLITICAL ECONOMY. 2d Ed., revised and enlarged. 8vo. 12s. 6d.
THE WAGES QUESTION. 8vo. 14s.
WICKSTEED.—ALPHABET OF ECONOMIC SCIENCE. By PHILIP H. WICK-
STEED, M.A. Part I. Elements of the Theory of Value or Worth. Gl. 8vo.
2s. 6d.

LAW AND POLITICS.

ADAMS AND CUNNINGHAM.—THE SWISS CONFEDERATION. By Sir
F. O. ADAMS and C. CUNNINGHAM. 8vo. 14s.
ANGLO-SAXON LAW, ESSAYS ON.—Contents : Anglo-Saxon Law Courts, Land
and Family Law, and Legal Procedure. 8vo. 18s
BALL.—THE STUDENT'S GUIDE TO THE BAR. By WALTER W. R. BALL, M.A.,
Fellow and Assistant Tutor of Trinity College, Cambridge. 4th Ed., revised.
Cr. 8vo. 2s. 6d.
BIGELOW.—HISTORY OF PROCEDURE IN ENGLAND FROM THE NORMAN
CONQUEST. The Norman Period, 1066-1204. By MELVILLE M. BIGELOW,
Ph.D., Harvard University. 8vo. 16s.
BRYCE.—THE AMERICAN COMMONWEALTH. By JAMES BRYCE, M.P., D.C.L.,
Regius Professor of Civil Law in the University of Oxford. Two Volumes.
Ex. cr. 8vo. 25s. Part I. The National Government. Part II. The State
Governments. Part III. The Party System. Part IV. Public Opinion.
Part V. Illustrations and Reflections. Part VI. Social Institutions.
BUCKLAND.—OUR NATIONAL INSTITUTIONS. A Short Sketch for Schools.
By ANNA BUCKLAND. With Glossary. 18mo. 1s.
DICEY.—INTRODUCTION TO THE STUDY OF THE LAW OF THE CONSTITU-
TION. By A. V. DICEY, B.C.L., Vinerian Professor of English Law in the
University of Oxford. 3d Ed. 8vo. 12s. 6d.
DILKE.—PROBLEMS OF GREATER BRITAIN. By the Right Hon. Sir
CHARLES WENTWORTH DILKE. With Maps. 2 vols. 8vo. 36s.
DONISTHORPE.—INDIVIDUALISM : A System of Politics. By WORDSWORTH
DONISTHORPE. 8vo. 14s.
ENGLISH CITIZEN, THE.—A Series of Short Books on his Rights and Responsi-
bilities. Edited by HENRY CRAIK. Cr. 8vo. 3s. 6d. each.
CENTRAL GOVERNMENT. By H. D. TRAILL, D.C.L.
THE ELECTORATE AND THE LEGISLATURE. By SPENCER WALPOLE.
THE POOR LAW. By Rev. T. W. FOWLE, M.A.
THE NATIONAL BUDGET ; THE NATIONAL DEBT ; TAXES AND RATES.
By A. J. WILSON.
THE STATE IN RELATION TO LABOUR. By W. STANLEY JEVONS, LL.D.
THE STATE AND THE CHURCH. By the Hon. ARTHUR ELLIOT, M.P.
FOREIGN RELATIONS. By SPENCER WALPOLE.
THE STATE IN ITS RELATION TO TRADE. By Sir T. H. FARRER, Bart.
LOCAL GOVERNMENT. By M. D. CHALMERS, M.A.
THE STATE IN ITS RELATION TO EDUCATION. By HENRY CRAIK, LL.D.
THE LAND LAWS. By Sir F. POLLOCK, Bart., Professor of Jurisprudence in
the University of Oxford.
COLONIES AND DEPENDENCIES. Part I. INDIA. By J. S. COTTON, M.A.
II. THE COLONIES. By E. J. PAYNE, M.A.
JUSTICE AND POLICE. By F. W. MAITLAND.
THE PUNISHMENT AND PREVENTION OF CRIME. By Colonel Sir EDMUND
DU CANE, K.C.B., Chairman of Comissioners of Prisons.

HOLMES.—THE COMMON LAW. By O. W. HOLMES Jun. Demy 8vo. 12s.

MAITLAND.—PLEAS OF THE CROWN FOR THE COUNTY OF GLOUCESTER BEFORE THE ABBOT OF READING AND HIS FELLOW JUSTICES ITINERANT, IN THE FIFTH YEAR OF THE REIGN OF KING HENRY THE THIRD, AND THE YEAR OF GRACE 1221. By F. W. MAITLAND. 8vo. 7s. 6d.

PATERSON.—Works by JAMES PATERSON, Barrister-at-Law.
COMMENTARIES ON THE LIBERTY OF THE SUBJECT, AND THE LAWS OF ENGLAND RELATING TO THE SECURITY OF THE PERSON. Cheaper Issue. Two Vols. Cr. 8vo. 21s.
THE LIBERTY OF THE PRESS, SPEECH, AND PUBLIC WORSHIP. Being Commentaries on the Liberty of the Subject and the Laws of England. Cr. 8vo. 12s.

PHILLIMORE.—PRIVATE LAW AMONG THE ROMANS. From the Pandects. By J. G. PHILLIMORE, Q.C. 8vo. 16s.

POLLOCK.—ESSAYS IN JURISPRUDENCE AND ETHICS. By Sir FREDERICK POLLOCK, Bart., Corpus Christi Professor of Jurisprudence in the University of Oxford. 8vo. 10s. 6d.
INTRODUCTION TO THE HISTORY OF THE SCIENCE OF POLITICS. By the same. Cr. 8vo. 2s. 6d.

RICHEY.—THE IRISH LAND LAWS. By ALEXANDER G. RICHEY, Q.C., Deputy Regius Professor of Feudal English Law in the University of Dublin. Cr. 8vo. 3s. 6d.

SIDGWICK.—THE ELEMENTS OF POLITICS. By HENRY SIDGWICK, LL.D. 8vo. [In the Press.

STEPHEN.—Works by Sir J. FITZJAMES STEPHEN, Q.C., K.C.S.I., a Judge of the High Court of Justice, Queen's Bench Division.
A DIGEST OF THE LAW OF EVIDENCE. 5th Ed., revised and enlarged. Cr. 8vo. 6s.
A DIGEST OF THE CRIMINAL LAW: CRIMES AND PUNISHMENTS. 4th Ed., revised. 8vo. 16s.
A DIGEST OF THE LAW OF CRIMINAL PROCEDURE IN INDICTABLE OFFENCES. By Sir J. F. STEPHEN, K.C.S.I., and H. STEPHEN, LL.M., of the Inner Temple, Barrister-at-Law. 8vo. 12s. 6d.
A HISTORY OF THE CRIMINAL LAW OF ENGLAND. Three Vols. 8vo. 48s.
GENERAL VIEW OF THE CRIMINAL LAW OF ENGLAND. 2d Ed. 8vo. 14s. The first edition of this work was published in 1863. The new edition is substantially a new work, intended as a text-book on the Criminal Law for University and other Students, adapted to the present day.

ANTHROPOLOGY.

FLOWER.—FASHION IN DEFORMITY, as Illustrated in the Customs of Barbarous and Civilised Races. By Prof. FLOWER, F.R.S. Illustrated. Cr. 8vo. 2s. 6d.

FRAZER.—THE GOLDEN BOUGH. A Study in Comparative Religion. By J. G. FRAZER, M.A., Fellow of Trinity College, Cambridge. 2 vols. 8vo. 28s.

M'LENNAN.—THE PATRIARCHAL THEORY. Based on the papers of the late JOHN F. M'LENNAN. Edited by DONALD M'LENNAN, M.A., Barrister-at-Law. 8vo. 14s.
STUDIES IN ANCIENT HISTORY. Comprising a Reprint of "Primitive Marriage." An inquiry into the origin of the form of capture in Marriage Ceremonies. 8vo. 16s.

TYLOR.—ANTHROPOLOGY. An Introduction to the Study of Man and Civilisation. By E. B. TYLOR, F.R.S. Illustrated. Cr. 8vo. 7s. 6d.

EDUCATION.

ARNOLD.—REPORTS ON ELEMENTARY SCHOOLS. 1852-1882. By MATTHEW ARNOLD, D.C.L. Edited by the Right Hon. Sir FRANCIS SANDFORD, K.C.B. Cheaper Issue. Cr. 8vo. 3s. 6d.

BALL.—THE STUDENT'S GUIDE TO THE BAR. By WALTER W. R. BALL, M.A., Fellow and Assistant Tutor of Trinity College, Cambridge. 4th Ed., revised. Cr. 8vo. 2s. 6d.

BLAKISTON.—THE TEACHER. Hints on School Management. A handbook for Managers, Teachers' Assistants, and Pupil Teachers. By J. R. BLAKISTON. Cr. 8vo. 2s. 6d. (Recommended by the London, Birmingham, and Leicester School Boards.)

CALDERWOOD.—ON TEACHING. By Prof. HENRY CALDERWOOD. New Ed. Ex. fcap. 8vo. 2s. 6d.

FITCH.—NOTES ON AMERICAN SCHOOLS AND TRAINING COLLEGES. Reprinted from the Report of the English Education Department for 1888-89, with permission of the Controller of H.M.'s Stationery Office. By J. G. FITCH, M.A. Gl. 8vo. 2s. 6d.

GEIKIE.—THE TEACHING OF GEOGRAPHY. A Practical Handbook for the use of Teachers. By ARCHIBALD GEIKIE, F.R.S., Director-General of the Geological Survey of the United Kingdom. Cr. 8vo. 2s.

GLADSTONE.—OBJECT TEACHING.—A Lecture delivered at the Pupil-Teacher Centre, William Street Board School, Hammersmith. By J. H. GLADSTONE, F.R.S. With an Appendix. Cr. 8vo. 3d.

SPELLING REFORM FROM A NATIONAL POINT OF VIEW. By the same. Cr. 8vo. 1s. 6d.

HERTEL.—OVERPRESSURE IN HIGH SCHOOLS IN DENMARK. By Dr. HERTEL. Translated by C. G. SÖRENSEN. With Introduction by Sir J. CRICHTON-BROWNE, F.R.S. Cr. 8vo. 3s. 6d.

TECHNICAL KNOWLEDGE.

(SEE ALSO MECHANICS, LAW, AND MEDICINE.)

Civil and Mechanical Engineering; Military and Naval Science;
Agriculture; Domestic Economy; Book-Keeping.

CIVIL AND MECHANICAL ENGINEERING.

ALEXANDER AND THOMSON.—ELEMENTARY APPLIED MECHANICS. By T. ALEXANDER, Professor of Civil Engineering, Trinity College, Dublin, and A. W. THOMSON, Lecturer in Engineering at the Technical College, Glasgow. Part II. TRANSVERSE STRESS. Cr. 8vo. 10s. 6d.

CHALMERS.—GRAPHICAL DETERMINATION OF FORCES IN ENGINEER-ING STRUCTURES. By J. B. CHALMERS, C.E. Illustrated. 8vo. 24s.

COTTERILL.—APPLIED MECHANICS: an Elementary General Introduction to the Theory of Structures and Machines. By J. H. COTTERILL, F.R.S., Professor of Applied Mechanics in the Royal Naval College, Greenwich. 2d Ed. 8vo. 18s.

COTTERILL AND SLADE.—ELEMENTARY MANUAL OF APPLIED MECHAN-ICS. By Prof. J. H. COTTERILL and J. H. SLADE. Cr. 8vo.   [*In the Press.*

KENNEDY.—THE MECHANICS OF MACHINERY. By A. B. W. KENNEDY, F.R.S. Illustrated. Cr. 8vo. 12s. 6d.

REULEAUX.—THE KINEMATICS OF MACHINERY. Outlines of a Theory of Machines. By Prof. F. REULEAUX. Translated and Edited by Prof. A. B. W. KENNEDY, F.R.S. Illustrated. 8vo. 21s.

WHITHAM.—STEAM-ENGINE DESIGN. For the Use of Mechanical Engineers, Students, and Draughtsmen. By J. M. WHITHAM, Professor of Engineering, Arkansas Industrial University. Illustrated. 8vo. 25s.

YOUNG.—SIMPLE PRACTICAL METHODS OF CALCULATING STRAINS ON GIRDERS, ARCHES, AND TRUSSES. With a Supplementary Essay on Economy in Suspension Bridges. By E. W. YOUNG, C.E. With Diagrams. 8vo. 7s. 6d.

MILITARY AND NAVAL SCIENCE.

AITKEN.—THE GROWTH OF THE RECRUIT AND YOUNG SOLDIER. With a view to the selection of "Growing Lads" for the Army, and a Regulated System of Training for Recruits. By Sir W. AITKEN, F.R.S., Professor of Pathology in the Army Medical School. Cr. 8vo. 8s. 6d.

ARMY PRELIMINARY EXAMINATION, 1882-1889, Specimens of Papers set at the. With Answers to the Mathematical Questions. Subjects: Arithmetic, Algebra, Euclid, Geometrical Drawing, Geography, French, English Dictation. Cr. 8vo. 3s. 6d.

MERCUR.—ELEMENTS OF THE ART OF WAR. Prepared for the use of Cadets of the United States Military Academy. By JAMES MERCUR, Professor of Civil Engineering at the United States Academy, West Point, New York. 2d Ed., revised and corrected. 8vo. 17s.

PALMER.—TEXT BOOK OF PRACTICAL LOGARITHMS AND TRIGONO-METRY.—By J. H. PALMER, Head Schoolmaster, R.N., H.M.S. *Cambridge*, Devonport. Gl. 8vo. 4s. 6d.

ROBINSON.—TREATISE ON MARINE SURVEYING. Prepared for the use of younger Naval Officers. With Questions for Examinations and Exercises principally from the Papers of the Royal Naval College. With the results. By Rev. JOHN L. ROBINSON, Chaplain and Instructor in the Royal Naval College, Greenwich. Illustrated. Cr. 8vo. 7s. 6d.

SANDHURST MATHEMATICAL PAPERS, for Admission into the Royal Military College, 1881-1889. Edited by E. J. BROOKSMITH, B.A., Instructor in Mathematics at the Royal Military Academy, Woolwich. Cr. 8vo. [*Immediately.*

SHORTLAND.—NAUTICAL SURVEYING. By the late Vice-Admiral SHORTLAND, LL.D. 8vo. 21s.

WILKINSON.—THE BRAIN OF AN ARMY. A Popular Account of the German General Staff. By SPENSER WILKINSON. Cr. 8vo. 2s. 6d.

WOLSELEY.—Works by General Viscount WOLSELEY, G.C.M.G.

THE SOLDIER'S POCKET-BOOK FOR FIELD SERVICE. 5th Ed., revised and enlarged. 16mo. Roan. 5s.

FIELD POCKET-BOOK FOR THE AUXILIARY FORCES. 16mo. 1s. 6d.

WOOLWICH MATHEMATICAL PAPERS, for Admission into the Royal Military Academy, Woolwich, 1880-1888 inclusive. Edited by E. J. BROOKSMITH, B.A., Instructor in Mathematics at the Royal Military Academy, Woolwich. Cr. 8vo. 6s.

AGRICULTURE.

FRANKLAND.—AGRICULTURAL CHEMICAL ANALYSIS, A Handbook of. By PERCY F. FRANKLAND, F.R.S., Professor of Chemistry, University College, Dundee. Founded upon *Leitfaden für die Agriculture Chemiche Analyse*, von Dr. F. KROCKER. Cr. 8vo. 7s. 6d.

SMITH.—DISEASES OF FIELD AND GARDEN CROPS, CHIEFLY SUCH AS ARE CAUSED BY FUNGI. By WORTHINGTON G. SMITH, F.L.S., Illustrated. Fcap. 8vo. 4s. 6d.

TANNER.—ELEMENTARY LESSONS IN THE SCIENCE OF AGRICULTURAL PRACTICE. By HENRY TANNER, F.C.S., M.R.A.C., Examiner in the Principles of Agriculture under the Government Department of Science. Fcap. 8vo. 3s. 6d.

FIRST PRINCIPLES OF AGRICULTURE. By the same. 18mo. 1s.

THE PRINCIPLES OF AGRICULTURE. By the same. A Series of Reading Books for use in Elementary Schools. Ex. fcap. 8vo.

  I. The Alphabet of the Principles of Agriculture. 6d.

  II. Further Steps in the Principles of Agriculture. 1s.

  III. Elementary School Readings on the Principles of Agriculture for the third stage. 1s.

WARD.—TIMBER AND SOME OF ITS DISEASES. By H. MARSHALL WARD, M.A., F.L.S., F.R.S., Fellow of Christ's College, Cambridge, Professor of Botany at the Royal Indian Engineering College, Cooper's Hill. With Illustrations. Cr. 8vo. 6s.

DOMESTIC ECONOMY.

BARKER.—FIRST LESSONS IN THE PRINCIPLES OF COOKING. By LADY BARKER. 18mo. 1s.

BERNERS.—FIRST LESSONS ON HEALTH. By J. BERNERS. 18mo. 1s.

BLYTH.—A MANUAL OF PUBLIC HEALTH. By A. WYNTER BLYTH, M.R.C.S. 8vo. [In the Press.

COOKERY BOOK.—THE MIDDLE CLASS COOKERY BOOK. Edited by the Manchester School of Domestic Cookery. Fcap. 8vo. 1s. 6d.

CRAVEN.—A GUIDE TO DISTRICT NURSES. By Mrs. DACRE CRAVEN (née FLORENCE SARAH LEES), Hon. Associate of the Order of St. John of Jerusalem, &c. Cr 8vo. 2s. 6d.

FREDERICK.—HINTS TO HOUSEWIVES ON SEVERAL POINTS, PARTICULARLY ON THE PREPARATION OF ECONOMICAL AND TASTEFUL DISHES. By Mrs. FREDERICK. Cr. 8vo. 1s.

GRAND'HOMME.—CUTTING-OUT AND DRESSMAKING. From the French of Mdlle. E. GRAND'HOMME. With Diagrams. 18mo. 1s.

JEX-BLAKE.—THE CARE OF INFANTS. A Manual for Mothers and Nurses. By SOPHIA JEX-BLAKE, M.D., Lecturer on Hygiene at the London School of Medicine for Women. 18mo. 1s.

RATHBONE.—THE HISTORY AND PROGRESS OF DISTRICT NURSING FROM ITS COMMENCEMENT IN THE YEAR 1859 TO THE PRESENT DATE, including the foundation by the Queen of the Queen Victoria Jubilee Institute for Nursing the Poor in their own Homes. By WILLIAM RATHBONE, M.P. Cr. 8vo. 2s. 6d.

TEGETMEIER.—HOUSEHOLD MANAGEMENT AND COOKERY. With an Appendix of Recipes used by the Teachers of the National School of Cookery. By W. B. TEGETMEIER. Compiled at the request of the School Board for London. 18mo. 1s.

WRIGHT.—THE SCHOOL COOKERY-BOOK. Compiled and Edited by C. E. GUTHRIE WRIGHT, Hon. Sec. to the Edinburgh School of Cookery. 18mo. 1s.

BOOK-KEEPING.

THORNTON.—FIRST LESSONS IN BOOK-KEEPING. By J. THORNTON. Cr. 8vo. 2s. 6d. KEY. Oblong 4to. 10s. 6d.

PRIMER OF BOOK-KEEPING. By the Same. 18mo. 1s. [Key Immediately.

GEOGRAPHY.

(SEE ALSO PHYSICAL GEOGRAPHY.)

BARTHOLOMEW.—THE ELEMENTARY SCHOOL ATLAS. By JOHN BARTHOLOMEW, F.R.G.S. 4to. 1s.

This Elementary Atlas is designed to illustrate the principal text-books on Elementary Geography.

PHYSICAL AND POLITICAL SCHOOL ATLAS, Consisting of 80 Maps and complete Index. By the Same. Prepared for the use of Senior Pupils. Royal 4to. [In the Press.

THE LIBRARY REFERENCE ATLAS OF THE WORLD. By the Same. A Complete Series of 84 Modern Maps. With Geographical Index to 100,000 places. Half-morocco. Gilt edges. Folio. £2:12:6 net.

₊ This work has been designed with the object of supplying the public with a thoroughly complete and accurate atlas of Modern Geography, in a convenient reference form, and at a moderate price.

CLARKE.—CLASS-BOOK OF GEOGRAPHY. By C. B. CLARKE, F.R.S. New Ed., revised 1889, with 18 Maps. Fcap. 8vo. Paper covers, 3s. Cloth, 3s. 6d.

GEIKIE.—Works by ARCHIBALD GEIKIE, F.R.S., Director-General of the Geological Survey of the United Kingdom.

THE TEACHING OF GEOGRAPHY. A Practical Handbook for the use of Teachers. Cr. 8vo. 2s.

GEOGRAPHY OF THE BRITISH ISLES. 18mo. 1s.

GREEN.—A SHORT GEOGRAPHY OF THE BRITISH ISLANDS. By JOHN RICHARD GREEN and A. S. GREEN. With Maps. Fcap. 8vo. 3s. 6d.

GROVE.—A PRIMER OF GEOGRAPHY. By Sir GEORGE GROVE, D.C.L. Illustrated. 18mo. 1s

KIEPERT.—A MANUAL OF ANCIENT GEOGRAPHY. By Dr. H. KIEPERT. Cr. 8vo. 5s.

MACMILLAN'S GEOGRAPHICAL SERIES.— Edited by ARCHIBALD GEIKIE, F.R.S., Director-General of the Geological Survey of the United Kingdom.

THE TEACHING OF GEOGRAPHY. A Practical Handbook for the Use of Teachers. By ARCHIBALD GEIKIE, F.R.S. Cr. 8vo. 2s.

MAPS AND MAP-MAKING. By W. A. ELDERTON. [In the Press.

GEOGRAPHY OF THE BRITISH ISLES. By A. GEIKIE, F.R.S. 18mo. 1s.

AN ELEMENTARY CLASS-BOOK OF GENERAL GEOGRAPHY. By H. R. MILL, D.Sc, Lecturer on Physiography and on Commercial Geography in the Heriot-Watt College, Edinburgh. Illustrated. Cr. 8vo. 3s. 6d.

GEOGRAPHY OF THE BRITISH COLONIES. By G. M. DAWSON and A. SUTHERLAND. [In preparation.

GEOGRAPHY OF EUROPE. By J. SIME, M.A. Illustrated. Gl. 8vo. [In the Press.

GEOGRAPHY OF INDIA. By H. F. BLANFORD, F.G.S. [In the Press.

GEOGRAPHY OF NORTH AMERICA. By Prof. N. S. SHALER. [In preparation.

ADVANCED CLASS-BOOK OF THE GEOGRAPHY OF BRITAIN.

₁ Other volumes will be announced in due course.

STRACHEY.—LECTURES ON GEOGRAPHY. By General RICHARD STRACHEY, R.E. Cr. 8vo. 4s. 6d.

HISTORY.

ARNOLD.—THE SECOND PUNIC WAR. Being Chapters from THE HISTORY OF ROME, by the late THOMAS ARNOLD, D.D., Headmaster of Rugby. Edited, with Notes, by W. T. ARNOLD, M.A. With 8 Maps. Cr. 8vo. 8s. 6d.

ARNOLD.—THE ROMAN SYSTEM OF PROVINCIAL ADMINISTRATION TO THE ACCESSION OF CONSTANTINE THE GREAT. By W. T. ARNOLD, M.A. Cr. 8vo. 6s.

BEESLY.—STORIES FROM THE HISTORY OF ROME. By Mrs. BEESLY. Fcap. 8vo. 2s. 6d.

BRYCE.—Works by JAMES BRYCE, M.P., D.C.L., Regius Professor of Civil Law in the University of Oxford.

THE HOLY ROMAN EMPIRE. 9th Ed. Cr. 8vo. 7s. 6d.
₁ Also a Library Edition. Demy 8vo. 14s.

THE AMERICAN COMMONWEALTH. 2 vols. Ex. cr. 8vo. 25s. Part I. The National Government. Part II. The State Governments. Part III. The Party System. Part IV. Public Opinion. Part V. Illustrations and Reflections. Part VI. Social Institutions.

BUCKLEY.—A HISTORY OF ENGLAND FOR BEGINNERS. By ARABELLA B. BUCKLEY. With Maps and Tables. Gl. 8vo. 3s.

BURY.—A HISTORY OF THE LATER ROMAN EMPIRE FROM ARCADIUS TO IRENE, A.D. 395-800. By JOHN B. BURY, M.A., Fellow of Trinity College, Dublin. 2 vols. 8vo. 32s.

ENGLISH STATESMEN, TWELVE. Cr. 8vo. 2s. 6d. each.
WILLIAM THE CONQUEROR. By EDWARD A. FREEMAN, D.C.L., LL.D.
HENRY II. By Mrs. J. R. GREEN.

EDWARD I. By F. YORK POWELL.   [*In preparation.*
HENRY VII. By JAMES GAIRDNER.
CARDINAL WOLSEY. By Professor M. CREIGHTON.
ELIZABETH. By E. S. BEESLY.   [*In preparation.*
OLIVER CROMWELL. By FREDERIC HARRISON.
WILLIAM III. By H. D. TRAILL.
WALPOLE. By JOHN MORLEY.
CHATHAM. By JOHN MORLEY   [*In preparation.*
PITT. By JOHN MORLEY.   [*In preparation.*
PEEL. By J. R. THURSFIELD.   [*In the Press.*

FISKE.—Works by JOHN FISKE, formerly Lecturer on Philosophy at Harvard University.

THE CRITICAL PERIOD IN AMERICAN HISTORY, 1783-1789. Ex. cr. 8vo. 10s. 6d.

THE BEGINNINGS OF NEW ENGLAND ; or, The Puritan Theocracy in its Relations to Civil and Religious Liberty. Cr. 8vo. 7s. 6d.

FREEMAN.—Works by EDWARD A. FREEMAN, D.C.L., Regius Professor of Modern History in the University of Oxford, &c.

OLD ENGLISH HISTORY. With Maps. Ex. fcap. 8vo. 6s.

A SCHOOL HISTORY OF ROME. Cr. 8vo.   [*In preparation.*

METHODS OF HISTORICAL STUDY. 8vo. 10s. 6d.

THE CHIEF PERIODS OF EUROPEAN HISTORY. Six Lectures. With an Essay on Greek Cities under Roman Rule. 8vo. 10s. 6d.

HISTORICAL ESSAYS. First Series. 4th Ed. 8vo. 10s. 6d.

HISTORICAL ESSAYS. Second Series. 3d Ed., with additional Essays. 8vo. 10s. 6d.

HISTORICAL ESSAYS. Third Series. 8vo. 12s.

THE GROWTH OF THE ENGLISH CONSTITUTION FROM THE EARLIEST TIMES. 4th Ed. Cr. 8vo. 5s.

GENERAL SKETCH OF EUROPEAN HISTORY. Enlarged, with Maps, etc. 18mo. 3s. 6d.

PRIMER OF EUROPEAN HISTORY. 18mo. 1s. (*History Primers.*)

FRIEDMANN.—ANNE BOLEYN. A Chapter of English History, 1527-1536. By PAUL FRIEDMANN. 2 vols. 8vo. 28s.

FYFFE.—A SCHOOL HISTORY OF GREECE. By C. A. FYFFE, M.A., late Fellow of University College, Oxford. Cr. 8vo.   [*In preparation.*

GREEN.—Works by JOHN RICHARD GREEN, LL.D., late Honorary Fellow of Jesus College, Oxford.

A SHORT HISTORY OF THE ENGLISH PEOPLE. New and Revised Ed. With Maps, Genealogical Tables, and Chronological Annals. Cr. 8vo. 8s. 6d. 150th Thousand.

Also the same in Four Parts. With the corresponding portion of Mr. Tait's "Analysis." Crown 8vo. 3s. each. Part I. 607-1265. Part II. 1204-1553. Part III. 1540-1689. Part IV. 1660-1873.

HISTORY OF THE ENGLISH PEOPLE. In four vols. 8vo. 16s. each.

Vol. I.—Early England, 449-1071 ; Foreign Kings, 1071-1214 ; The Charter, 1214-1291 ; The Parliament, 1307-1461. With 8 Maps.

Vol. II.—The Monarchy, 1461-1540 ; The Reformation, 1540-1603.

Vol. III.—Puritan England, 1603-1660 ; The Revolution, 1660-1688. With four Maps.

Vol. IV.—The Revolution, 1688-1760 ; Modern England, 1760-1815. With Maps and Index.

THE MAKING OF ENGLAND. With Maps. 8vo. 16s.

THE CONQUEST OF ENGLAND. With Maps and Portrait. 8vo. 18s.

ANALYSIS OF ENGLISH HISTORY, based on Green's "Short History of the English People." By C. W. A. TAIT, M.A., Assistant Master at Clifton College. Crown 8vo. 3s. 6d.

READINGS FROM ENGLISH HISTORY. Selected and Edited by JOHN RICHARD GREEN. Three Parts. Gl. 8vo. 1s. 6d. each. I. Hengist to Cressy. II. Cressy to Cromwell. III. Cromwell to Balaklava.

GUEST.—LECTURES ON THE HISTORY OF ENGLAND. By M. J. GUEST. With Maps. Cr. 8vo. 6s.

HISTORICAL COURSE FOR SCHOOLS.—Edited by E. A. FREEMAN, D.C.L., Regius Professor of Modern History in the University of Oxford. 18mo.

GENERAL SKETCH OF EUROPEAN HISTORY. By E. A. FREEMAN, D.C.L. New Ed., revised and enlarged. With Chronological Table, Maps, and Index. 3s. 6d.

HISTORY OF ENGLAND. By EDITH THOMPSON. New Ed., revised and enlarged. With Coloured Maps. 2s. 6d.

HISTORY OF SCOTLAND. By MARGARET MACARTHUR. 2s.

HISTORY OF ITALY. By Rev. W. HUNT, M.A. New Ed. With Coloured Maps. 3s. 6d.

HISTORY OF GERMANY. By J. SIME, M.A. New Ed., revised. 3s.

HISTORY OF AMERICA. By JOHN A. DOYLE. With Maps. 4s. 6d.

HISTORY OF EUROPEAN COLONIES. By E. J. PAYNE, M.A. With Maps. 4s. 6d.

HISTORY OF FRANCE. By CHARLOTTE M. YONGE. With Maps. 3s. 6d.

HISTORY OF GREECE. By EDWARD A. FREEMAN, D.C.L. [In preparation.

HISTORY OF ROME. By EDWARD A. FREEMAN, D.C.L. [In preparation.

HISTORY PRIMERS.—Edited by JOHN RICHARD GREEN, LL.D. 18mo. 1s. each.

ROME. By Rev. M. CREIGHTON, M.A., Dixie Professor of Ecclesiastical History in the University of Cambridge. Maps.

GREECE. By C. A. FYFFE, M.A., late Fellow of University College, Oxford. Maps.

EUROPE. By E. A. FREEMAN, D.C.L. Maps.

FRANCE. By CHARLOTTE M. YONGE.

GREEK ANTIQUITIES. By Rev. J. P. MAHAFFY, D.D. Illustrated.

CLASSICAL GEOGRAPHY. By H. F. TOZER, M.A.

GEOGRAPHY. By Sir G. GROVE, D.C.L. Maps.

ROMAN ANTIQUITIES. By Prof. WILKINS, Litt.D. Illustrated.

HOLE.—A GENEALOGICAL STEMMA OF THE KINGS OF ENGLAND AND FRANCE. By Rev. C. HOLE. On Sheet. 1s.

JENNINGS.—CHRONOLOGICAL TABLES. A synchronistic arrangement of the events of Ancient History (with an Index). By Rev. ARTHUR C. JENNINGS. 8vo. 5s.

LABBERTON.—NEW HISTORICAL ATLAS AND GENERAL HISTORY. By R. H. LABBERTON. 4to. New Ed., revised and enlarged. 15s.

LETHBRIDGE.—A SHORT MANUAL OF THE HISTORY OF INDIA. With an Account of INDIA AS IT IS. The Soil, Climate, and Productions; the People, their Races, Religions, Public Works, and Industries; the Civil Services, and System of Administration. By Sir ROPER LETHBRIDGE, Fellow of the Calcutta University. With Maps. Cr. 8vo. 5s.

MAHAFFY.—GREEK LIFE AND THOUGHT FROM THE AGE OF ALEX-ANDER TO THE ROMAN CONQUEST. By Rev. J. P. MAHAFFY, D.D., Fellow of Trinity College, Dublin. Cr. 8vo. 12s. 6d.

THE GREEK WORLD UNDER ROMAN SWAY. From Plutarch to Polybius. By the same Author. Cr. 8vo. [In the Press.

MARRIOTT.—THE MAKERS OF MODERN ITALY: MAZZINI, CAVOUR, GARI-BALDI. Three Lectures. By J. A. R. MARRIOTT, M.A., Lecturer in Modern History and Political Economy, Oxford. Cr. 8vo. 1s. 6d.

MICHELET.—A SUMMARY OF MODERN HISTORY. Translated by M. C. M. SIMPSON. Gl. 8vo. 4s. 6d.

NORGATE.—ENGLAND UNDER THE ANGEVIN KINGS. By KATE NORGATE. With Maps and Plans. 2 vols. 8vo. 32s.

OTTÉ.—SCANDINAVIAN HISTORY. By E. C. Otté. With Maps. Gl. 8vo. 6s.

SEELEY.—Works by J. R. Seeley, M.A., Regius Professor of Modern History in the University of Cambridge.
THE EXPANSION OF ENGLAND. Crown 8vo. 4s. 6d.
OUR COLONIAL EXPANSION. Extracts from the above. Cr. 8vo. Sewed. 1s.

TAIT.—ANALYSIS OF ENGLISH HISTORY, based on Green's "Short History of the English People." By C. W. A. Tait, M.A., Assistant Master at Clifton. Cr. 8vo. 3s. 6d.

WHEELER.—Works by J. Talboys Wheeler.
A PRIMER OF INDIAN HISTORY. Asiatic and European. 18mo. 1s.
COLLEGE HISTORY OF INDIA, ASIATIC AND EUROPEAN. With Maps. Cr. 8vo. 3s. 6d.
A SHORT HISTORY OF INDIA AND OF THE FRONTIER STATES OF AFGHANISTAN, NEPAUL, AND BURMA. With Maps. Cr. 8vo. 12s.

YONGE.—Works by Charlotte M. Yonge.
CAMEOS FROM ENGLISH HISTORY. Ex. fcap. 8vo. 5s. each. (1) FROM ROLLO TO EDWARD II. (2) THE WARS IN FRANCE. (3) THE WARS OF THE ROSES. (4) REFORMATION TIMES. (5) ENGLAND AND SPAIN. (6) FORTY YEARS OF STUART RULE (1603-1643). (7) REBELLION AND RESTORATION (1642-1678.)
EUROPEAN HISTORY. Narrated in a Series of Historical Selections from the Best Authorities. Edited and arranged by E. M. Sewell and C. M. Yonge. Cr. 8vo. First Series, 1003-1154. 6s. Second Series, 1088-1228. 6s.
THE VICTORIAN HALF CENTURY—A JUBILEE BOOK. With a New Portrait of the Queen. Cr. 8vo., paper covers, 1s. Cloth, 1s. 6d.

ART.

ANDERSON.—LINEAR PERSPECTIVE AND MODEL DRAWING. A School and Art Class Manual, with Questions and Exercises for Examination, and Examples of Examination Papers. By Laurence Anderson. Illustrated. 8vo. 2s.

COLLIER.—A PRIMER OF ART. By the Hon. John Collier. Illustrated. 18mo. 1s.

COOK.—THE NATIONAL GALLERY: A POPULAR HANDBOOK TO. By Edward T. Cook, with a preface by John Ruskin, LL.D., and Selections from his Writings. 3d Ed. Cr. 8vo. Half Morocco, 14s.
₊ Also an Edition on large paper, limited to 250 copies. 2 vols. 8vo.

DELAMOTTE.—A BEGINNER'S DRAWING BOOK. By P. H. Delamotte, F.S.A. Progressively arranged. New Ed., improved. Cr. 8vo. 3s. 6d.

ELLIS.—SKETCHING FROM NATURE. A Handbook for Students and Amateurs. By Tristram J. Ellis. Illustrated by H. Stacy Marks, R.A., and the Author. New Ed., revised and enlarged. Cr. 8vo. 3s. 6d.

GROVE.—A DICTIONARY OF MUSIC AND MUSICIANS. A.D. 1450—1889. Edited by Sir George Grove, D.C.L. In four vols. 8vo. Price 21s. each. Also in Parts.
Parts I. to XIV., Parts XIX.—XXII., 3s. 6d. each. Parts XV., XVI., 7s. Parts XVII., XVIII., 7s. Parts XXIII.—XXV. (Appendix), 9s.
A COMPLETE INDEX TO THE ABOVE. By Mrs. E. Wodehouse. 8vo. 7s. 6d.

HUNT.—TALKS ABOUT ART. By William Hunt. With a Letter from Sir J. E. Millais, Bart., R.A. Cr. 8vo. 3s. 6d.

MELDOLA.—THE CHEMISTRY OF PHOTOGRAPHY. By Raphael Meldola, F.R.S., Professor of Chemistry in the Technical College, Finsbury. Cr. 8vo. 6s.

TAYLOR.—A PRIMER OF PIANOFORTE PLAYING. By Franklin Taylor. Edited by Sir George Grove. 18mo. 1s.

TAYLOR.—A SYSTEM OF SIGHT-SINGING FROM THE ESTABLISHED MUSICAL NOTATION ; based on the Principle of Tonic Relation, and Illustrated by Extracts from the Works of the Great Masters. By SEDLEY TAYLOR. 8vo. [In the Press.

TYRWHITT.—OUR SKETCHING CLUB. Letters and Studies on Landscape Art. By Rev. R. ST. JOHN TYRWHITT. With an authorised Reproduction of the Lessons and Woodcuts in Prof. Ruskin's "Elements of Drawing." 4th Ed. Cr. 8vo. 7s. 6d.

DIVINITY.

ABBOTT.—BIBLE LESSONS. By Rev. EDWIN A. ABBOTT, D.D. Cr. 8vo. 4s. 6d.

ABBOTT—RUSHBROOKE.—THE COMMON TRADITION OF THE SYNOPTIC GOSPELS, in the Text of the Revised Version. By Rev. EDWIN A. ABBOTT, D.D., and W. G. RUSHBROOKE, M.L. Cr. 8vo. 3s. 6d.

ARNOLD.—Works by MATTHEW ARNOLD.

A BIBLE-READING FOR SCHOOLS,—THE GREAT PROPHECY OF ISRAEL'S RESTORATION (Isaiah, Chapters xl.-lxvi.) Arranged and Edited for Young Learners. 18mo. 1s.

ISAIAH XL.—LXVI. With the Shorter Prophecies allied to it. Arranged and Edited, with Notes. Cr. 8vo. 5s.

ISAIAH OF JERUSALEM, IN THE AUTHORISED ENGLISH VERSION. With Introduction, Corrections and Notes. Cr. 8vo. 4s. 6d.

BENHAM.—A COMPANION TO THE LECTIONARY. Being a Commentary on the Proper Lessons for Sundays and Holy Days. By Rev. W. BENHAM, B.D. Cr. 8vo. 4s. 6d.

CASSEL.—MANUAL OF JEWISH HISTORY AND LITERATURE ; preceded by a BRIEF SUMMARY OF BIBLE HISTORY. By Dr. D. CASSEL. Translated by Mrs. H. LUCAS. Fcap. 8vo. 2s. 6d.

CROSS.—BIBLE READINGS SELECTED FROM THE PENTATEUCH AND THE BOOK OF JOSHUA. By Rev. JOHN A. CROSS. 2d Ed., enlarged, with Notes. Gl. 8vo. 2s. 6d.

DRUMMOND.—THE STUDY OF THEOLOGY, INTRODUCTION TO. By JAMES DRUMMOND, LL.D., Professor of Theology in Manchester New College, London. Cr. 8vo. 5s.

FARRAR.—Works by the Venerable Archdeacon F. W. FARRAR, D.D., F.R.S., Archdeacon and Canon of Westminster.

THE HISTORY OF INTERPRETATION. Being the Bampton Lectures, 1885. 8vo. 16s.

THE MESSAGES OF THE BOOKS. Being Discourses and Notes on the Books of the New Testament. 8vo. 14s.

GASKOIN.—THE CHILDREN'S TREASURY OF BIBLE STORIES. By Mrs. HERMAN GASKOIN. Edited with Preface by Rev. G. F. MACLEAR, D.D. 18mo. 1s. each. Part I.—OLD TESTAMENT HISTORY. Part II.—NEW TESTAMENT. Part III.—THE APOSTLES : ST. JAMES THE GREAT, ST. PAUL, AND ST. JOHN THE DIVINE.

GOLDEN TREASURY PSALTER.—Students' Edition. Being an Edition of "The Psalms Chronologically arranged, by Four Friends," with briefer Notes. 18mo. 3s. 6d.

GREEK TESTAMENT.—Edited, with Introduction and Appendices, by Bishop WESTCOTT and Dr. F. J. A. HORT. Two Vols. Cr. 8vo. 10s. 6d. each. Vol. I. The Text. Vol. II. Introduction and Appendix.

SCHOOL EDITION OF TEXT. 12mo, cloth. 4s. 6d. 18mo, morcceo, gilt edges. 6s. 6d.

GREEK TESTAMENT, SCHOOL READINGS IN THE. Being the outline of the life of our Lord, as given by St. Mark, with additions from the Text of the other Evangelists. Arranged and Edited, with Notes and Vocabulary, by Rev. A. CALVERT, M.A. Fcap. 8vo. 4s. 6d.

THE GOSPEL ACCORDING TO ST. MATTHEW. Being the Greek Text as revised by Bishop WESTCOTT and Dr. HORT. With Introduction and Notes by Rev. A. SLOMAN, M.A., Headmaster of Birkenhead School. Fcap. 8vo. 2s. 6d.

THE GOSPEL ACCORDING TO ST. MARK. Being the Greek Text as revised by Bishop WESTCOTT and Dr. HORT. With Introduction and Notes by Rev. J. O. F. MURRAY, M.A., Lecturer at Emmanuel College, Cambridge. Fcap. 8vo. [*In preparation.*]

THE GOSPEL ACCORDING TO ST. LUKE. Being the Greek Text as revised by Bishop WESTCOTT and Dr. HORT. With Introduction and Notes by Rev. JOHN BOND, M.A. [*In preparation.*]

THE ACTS OF THE APOSTLES. Being the Greek Text as revised by Bishop WESTCOTT and Dr. HORT. With Explanatory Notes by T. E. PAGE, M.A., Assistant Master at the Charterhouse. Fcap. 8vo. 4s. 6d.

GWATKIN.—CHURCH HISTORY TO THE BEGINNING OF THE MIDDLE AGES. By H. M. GWATKIN, M.A. 8vo. [*In preparation.*]

HARDWICK.—Works by Archdeacon HARDWICK.

A HISTORY OF THE CHRISTIAN CHURCH. Middle Age. From Gregory the Great to the Excommunication of Luther. Edited by W. STUBBS, D.D., Bishop of Oxford. With 4 Maps. Cr. 8vo. 10s. 6d.

A HISTORY OF THE CHRISTIAN CHURCH DURING THE REFORMATION. 9th. Ed. Edited by Bishop STUBBS. Cr. 8vo. 10s. 6d.

HOOLE.—THE CLASSICAL ELEMENT IN THE NEW TESTAMENT. Considered as a proof of its Genuineness, with an Appendix on the Oldest Authorities used in the Formation of the Canon. By CHARLES H. HOOLE, M.A., Student of Christ Church, Oxford. 8vo. 10s. 6d.

JENNINGS AND LOWE.—THE PSALMS, WITH INTRODUCTIONS AND CRITICAL NOTES. By A. C. JENNINGS, M.A. ; assisted in parts by W. H. LOWE, M.A. In 2 vols. 2d Ed., revised. Cr. 8vo. 10s. 6d. each.

KIRKPATRICK.—THE MINOR PROPHETS. Warburtonian Lectures. By Rev. Prof. KIRKPATRICK. [*In preparation.*]

KUENEN.—PENTATEUCH AND BOOK OF JOSHUA: an Historico - Critical Inquiry into the Origin and Composition of the Hexateuch. By A. KUENEN. Translated by P. H. WICKSTEED, M.A. 8vo. 14s.

LIGHTFOOT.—Works by the Right Rev. J. B. LIGHTFOOT, D.D., late Bishop of Durham.

ST. PAUL'S EPISTLE TO THE GALATIANS. A Revised Text, with Introduction, Notes, and Dissertations. 9th Ed., revised. 8vo. 12s.

ST. PAUL'S EPISTLE TO THE PHILIPPIANS. A Revised Text, with Introduction, Notes, and Dissertations. 9th. Ed., revised. 8vo. 12s.

ST. CLEMENT OF ROME—THE TWO EPISTLES TO THE CORINTHIANS. A Revised Text, with Introduction and Notes. 2 Vols. 8vo. [*In the Press.*]

ST. PAUL'S EPISTLES TO THE COLOSSIANS AND TO PHILEMON. A Revised Text, with Introductions, Notes, and Dissertations. 8th Ed., revised. 8vo. 12s.

THE APOSTOLIC FATHERS. Part II. ST. IGNATIUS—ST. POLYCARP. Revised Texts. With Introductions, Notes, Dissertations, and Translations. 2d Ed. 3 vols. 8vo. 48s.

THE APOSTOLIC FATHERS. Abridged Edition. With short Introductions, Greek Text, and English Translation. 8vo. [*In the Press.*]

ESSAYS ON THE WORK ENTITLED "SUPERNATURAL RELIGION." (Reprinted from the *Contemporary Review*). 8vo. 10s. 6d.

MACLEAR.—Works by the Rev. G. F. MACLEAR, D.D., Warden of St. Augustine's College, Canterbury.

ELEMENTARY THEOLOGICAL CLASS-BOOKS.

A SHILLING BOOK OF OLD TESTAMENT HISTORY. With Map. 18mo.

A SHILLING BOOK OF NEW TESTAMENT HISTORY. With Map. 18mo. These works have been carefully abridged from the Author's large manuals.

A CLASS-BOOK OF OLD TESTAMENT HISTORY. Maps. 18mo. 4s. 6d.

A CLASS BOOK OF NEW TESTAMENT HISTORY, including the Connection of the Old and New Testaments. With maps. 18mo. 5s. 6d.

AN INTRODUCTION TO THE THIRTY-NINE ARTICLES. 18mo.

AN INTRODUCTION TO THE CREEDS. 18mo. 2s. 6d. . [In the Press.

A CLASS-BOOK OF THE CATECHISM OF THE CHURCH OF ENGLAND. 18mo. 1s. 6d.

A FIRST CLASS-BOOK OF THE CATECHISM OF THE CHURCH OF ENGLAND. With Scripture Proofs. 18mo. 6d.

A MANUAL OF INSTRUCTION FOR CONFIRMATION AND FIRST COMMUNION. WITH PRAYERS AND DEVOTIONS. 32mo. 2s.

MAURICE.—THE LORD'S PRAYER, THE CREED, AND THE COMMANDMENTS. To which is added the Order of the Scriptures. By Rev. F. D. MAURICE, M.A. 18mo. 1s.

THE PENTATEUCH AND BOOK OF JOSHUA: an Historico-Critical Inquiry into the Origin and Composition of the Hexateuch. By A. KUENEN, Professor of Theology at Leiden. Translated by P. H. WICKSTEED, M.A. 8vo. 14s.

PROCTER.—A HISTORY OF THE BOOK OF COMMON PRAYER, with a Rationale of its Offices. By Rev. F. PROCTER. 18th Ed., revised and enlarged. Cr. 8vo. 10s. 6d.

PROCTER AND MACLEAR.—AN ELEMENTARY INTRODUCTION TO THE BOOK OF COMMON PRAYER. Re-arranged and supplemented by an Explanation of the Morning and Evening Prayer and the Litany. By Rev. F. PROCTER and Rev. Dr. MACLEAR. New and enlarged Edition, containing the Communion Service and the Confirmation and Baptismal Offices. 18mo. 2s. 6d.

THE PSALMS, WITH INTRODUCTIONS AND CRITICAL NOTES. By A. C. JENNINGS, M.A., Jesus College, Cambridge; assisted in parts by W. H. LOWE. M.A., Hebrew Lecturer at Christ's College, Cambridge. In 2 vols. 2d Ed., revised. Cr. 8vo. 10s. 6d. each.

RAMSAY.—THE CATECHISER'S MANUAL; or, the Church Catechism Illustrated and Explained. By Rev. ARTHUR RAMSAY. 18mo. 1s. 6d.

RYLE.—AN INTRODUCTION TO THE CANON OF THE OLD TESTAMENT. By Rev. H. E. RYLE, Hulsean Professor of Divinity in the University of Cambridge. Cr. 8vo.   [In preparation.

SIMPSON.—AN EPITOME OF THE HISTORY OF THE CHRISTIAN CHURCH DURING THE FIRST THREE CENTURIES, AND OF THE REFORMATION IN ENGLAND. By Rev. WILLIAM SIMPSON. 7th Ed. Fcap. 8vo. 3s. 6d.

ST. JAMES' EPISTLE.—The Greek Text with Introduction and Notes. By Rev. JOSEPH MAYOR, M.A., Professor of Moral Philosophy in King's College, London. 8vo.   [In the Press.

ST. JOHN'S EPISTLES.—The Greek Text, with Notes and Essays. By Right Rev. B. F. WESTCOTT, D.D., Bishop of Durham. 2d Ed., revised. 8vo. 12s. 6d.

ST. PAUL'S EPISTLES.—THE EPISTLE TO THE ROMANS. Edited by the Very Rev. C. J. VAUGHAN, D.D., Dean of Llandaff. 5th Ed. Cr. 8vo. 7s. 6d.

THE TWO EPISTLES TO THE CORINTHIANS, A COMMENTARY ON. By the late Rev. W. KAY, D.D., Rector of Great Leghs, Essex. 8vo. 9s.

THE EPISTLE TO THE GALATIANS. Edited by the Right Rev. J. B. LIGHTFOOT, D.D. 9th Ed. 8vo. 12s.

THE EPISTLES TO THE EPHESIANS, THE COLOSSIANS, AND PHILEMON; with Introductions and Notes, and an Essay on the Traces of Foreign Elements in the Theology of these Epistles. By Rev. J. LLEWELYN DAVIES, M.A. 8vo. 7s. 6d.

THE EPISTLE TO THE PHILIPPIANS. By the Right Rev. J. B. LIGHTFOOT, D.D. 9th Ed. 8vo. 12s.

THE EPISTLE TO THE PHILIPPIANS, with Translation, Paraphrase, and Notes for English Readers. By the Very Rev. C. J. VAUGHAN, D.D. Cr. 8vo. 5s.

THE EPISTLE TO THE COLOSSIANS AND TO PHILEMON. By the Right Rev. J. B. LIGHTFOOT, D.D. 8th Ed. 8vo. 12s.

THE EPISTLE TO THE THESSALONIANS, COMMENTARY ON THE GREEK TEXT. By JOHN EADIE, D.D. Edited by Rev. W. YOUNG, M.A., with Preface by Prof. CAIRNS. 8vo. 12s.

THE EPISTLE TO THE HEBREWS.—In Greek and English. With Critical and Explanatory Notes. Edited by Rev. F. RENDALL, M.A. Cr. 8vo. 6s.
THE ENGLISH TEXT, WITH COMMENTARY. By the same Editor. Cr. 8vo. 7s. 6d.
THE GREEK TEXT. With Notes by C. J. VAUGHAN, D.D., Dean of Llandaff. Cr. 8vo. 7s. 6d.
THE GREEK TEXT. With Notes and Essays by Bishop WESTCOTT, D.D. 8vo. 14s.

WESTCOTT.—Works by the Right Rev. BROOKE FOSS WESTCOTT, D.D., Bishop of Durham.
A GENERAL SURVEY OF THE HISTORY OF THE CANON OF THE NEW TESTAMENT DURING THE FIRST FOUR CENTURIES. 6th Ed. With Preface on "Supernatural Religion." Cr. 8vo. 10s. 6d.
INTRODUCTION TO THE STUDY OF THE FOUR GOSPELS. 7th Ed. Cr. 8vo. 10s. 6d.
THE BIBLE IN THE CHURCH. A Popular Account of the Collection and Reception of the Holy Scriptures in the Christian Churches. 18mo. 4s. 6d.
THE EPISTLES OF ST. JOHN. The Greek Text, with Notes and Essays. 2d Ed., revised. 8vo. 12s. 6d.
THE EPISTLE TO THE HEBREWS. The Greek Text, with Notes and Essays. 8vo. 14s.
SOME THOUGHTS FROM THE ORDINAL. Cr. 8vo. 1s. 6d.

WESTCOTT AND HORT.—THE NEW TESTAMENT IN THE ORIGINAL GREEK. The Text, revised by the Right Rev. Bishop WESTCOTT and Dr. F. J. A. HORT. 2 vols. Cr. 8vo. 10s. 6d. each. Vol. I. Text. Vol. II. Introduction and Appendix.
SCHOOL EDITION OF TEXT. 12mo. 4s. 6d. 18mo. Morocco, gilt edges. 6s. 6d.

WILSON.—THE BIBLE STUDENT'S GUIDE to the more Correct Understanding of the English Translation of the Old Testament, by reference to the original Hebrew. By WILLIAM WILSON, D.D., Canon of Winchester. 2d Ed., carefully revised. 4to. 25s.

WRIGHT.—THE COMPOSITION OF THE FOUR GOSPELS. A Critical Enquiry. By Rev. ARTHUR WRIGHT., M.A., Fellow and Tutor of Queen's College, Cambridge. Cr. 8vo. 5s.

WRIGHT.—THE BIBLE WORD-BOOK: A Glossary of Archaic Words and Phrases in the Authorised Version of the Bible and the Book of Common Prayer. By W. ALDIS WRIGHT, M.A., Vice-Master of Trinity College, Cambridge. 2d Ed., revised and enlarged. Cr. 8vo. 7s. 6d.

YONGE.—SCRIPTURE READINGS FOR SCHOOLS AND FAMILIES. By CHARLOTTE M. YONGE. In Five Vols. Ex. fcp. 8vo. 1s. 6d. each. With Comments. 3s. 6d. each.
FIRST SERIES.—GENESIS TO DEUTERONOMY. SECOND SERIES.—From JOSHUA to SOLOMON. THIRD SERIES.—The KINGS and the PROPHETS. FOURTH SERIES. —The GOSPEL TIMES. FIFTH SERIES.—APOSTOLIC TIMES.

ZECHARIAH—THE HEBREW STUDENT'S COMMENTARY ON ZECHARIAH, HEBREW AND LXX. With Excursus on Syllable-dividing, Metheg, Initial Dagesh, and Siman Rapheh. By W. H. LOWE, M.A., Hebrew Lecturer at Christ's College, Cambridge. 8vo. 10s. 6d.

Printed by R. & R. CLARK, *Edinburgh.*

778110

Printed in Great Britain by
Amazon.co.uk, Ltd.,
Marston Gate.